LAWS OF PHYSICAL SCIENCE

A REFERENCE BOOK

EDWIN FITCH NORTHRUP

GIFT *Certificate*

TO:

FROM:

DATE:

LAWS OF PHYSICAL SCIENCE

A REFERENCE BOOK

BY

EDWIN FITCH NORTHRUP

Ph.D.

Edwin Fitch Northrup (February 23, 1866 – May 13, 1940) was a professor of physics at Princeton University from 1910 to 1920.

He studied at Amherst College and the Johns Hopkins University, where he gained his Ph.D. in physics in 1895.

DIAMOND™ BOOKS

www.diamondbooks.ca

TORONTO, CANADA – 2017

DIAMOND BOOKS - CANADA

Toronto, ON, CANADA
http://www.diamondbooks.ca

BIBLIOGRAPHIC INFORMATION

' LAWS OF PHYSICAL SCIENCE - A Reference Book '
was first published in 1917
by: J. B. LIPPINCOTT COMPANY.
The Washington Square Press, Philadelphia, U. S. A.

PUBLISHED IN CANADA

Published in Canada by DIAMOND BOOKS - CANADA, an imprint of DIAMOND PUBLISHERS - http://www.diamondpublishers.com

Published Edition: July, 2017.

PAPERBACK EDITION : ISBN: 978-1-988357-84-3
E-BOOK EDITION : ISBN: 978-1-988357-85-0

PRINTED IN CANADA

PREFACE

Exact knowledge consists of accumulated facts and sets of formulated propositions respecting facts. Data, Mathematical relations and Physical laws constitute the three firm supports of Physical science and Engineering.

The *data* of physical science are readily accessible in several published tables of physical constants. The *mathematics* used in physical science has been summarized, classified and formulated, for ready reference, in many published books. The author is not aware, however, of any hand-book or reference work which contains a full list of the *general propositions* or *laws of science.*

Such reference lists are not without value, and this book has been prepared to fill an obvious gap in the literature of Physical Science. Furthermore, it appears to the author that students in any of the branches of Natural Science will not only find guidance, but will also derive inspiration by having before them under a single view the very epitome of the world's heritage of the fundamentals of its knowledge and wisdom. None will question that the fundamentals of science are its laws, principles, theorems and precise statements of the general properties of matter; but it is not always easy for students in one branch of science to find and to know the literature on important principles and facts in an entirely different, or even in closely allied branches of science. The author hopes that what has been here gathered together and classified will help such students in their search and give them the means to broaden their view.

We have chosen for a title, "Laws of Physical Science" but many general propositions, theorems and mere statements of important facts have been included which perhaps, if strictly considered, could not be discriminated as laws. Indeed, it was found impossible, in many cases, to decide if certain propositions possess sufficient generality and validity to deserve the title "law." When, however, such doubts existed, a policy of inclusion has been followed in preference to one of exclusion.

For convenience and system the general statements (in all 480 with title) have been classified in six sections: I—Mechanics; II—Hydrostatics, Hydrodynamics and Capillarity; III—Sound; IV—Heat and Physical Chemistry; V—Electricity and Magnetism; VI—Light.

Each law, proposition or general statement is characterized by giving it a heading or title. Each proposition covered by a title is followed by one, and in many cases by several references to easily accessible text-books, standard treatises, and, in a few cases, to original articles or papers, where one may find the propositions stated in different forms and additional information concerning them of authoritative character.

While many laws of Physical Science have had their origin with individual investigators, the perfected form of statement they now possess has been in the main reached by a process of intellectual growth in which many have taken part. It has seemed, therefore, wiser to make most of the references to treatises and text-books on physics, physical chemistry and chemistry, rather than to papers written by the authors of the propositions. Moreover, original papers, beside containing much extraneous matter, are not usually readily accessible as are text-books and treatises.

An alphabetically arranged bibliography of all books and journals referred to and a very full index, with duplicated references, to aid in the quick location of subject matter and proper names, are included.

The author expresses his acknowledgment to Mr. H. A. Frederick for assistance in collecting some of the material used, and to Prof. K. T. Compton for his careful reading of the manuscript. He further acknowledges with gratitude the unfailing and valuable assistance of his wife, Margaret Stewart Northrup, in collecting material and in arranging it for publication.

PALMER PHYSICAL LABORATORY
PRINCETON, N. J.
JANUARY, 1917.

THE AUTHOR

CONTENTS

I

MECHANICS

LAWS OF PHYSICAL SCIENCE

MECHANICS

NEWTON'S FIRST LAW OF MOTION.

Every body continues in its state of rest or of uniform motion in a straight line, except in so far as it may be compelled by force to change that state.

(Thomson and Tait, *Treatise on Natural Philosophy,* Part I, *art.* 244.)

NEWTON'S SECOND LAW OF MOTION.

Change of motion is proportional to force applied, and takes place in the direction of the straight line in which the force acts.

(Thomson and Tait, *Treatise on Natural Philosophy,* Part I, *art.* 251.)

NEWTON'S THIRD LAW OF MOTION.

To every action there is always an equal and contrary reaction; or, the mutual actions of any two bodies are always equal and oppositely directed.

(Thomson and Tait, *Treatise on Natural Philosophy,* Part I, *art.* 261.)

NEWTON'S LAW OF UNIVERSAL GRAVITATION.

All bodies attract each other with a force proportional to the product of their masses and inversely proportional to the square of the distance between them.

Thus, $F = G \frac{M M'}{r^2}$.

G is called the Newtonian constant.

Cavendish (1797–1798) first measured G experimentally. In 1874 C. V. Boys obtained $6.657_6 \times 10^{-8}$ on the C.G.S. system for the value of G.

(Consult *Scientific Memoirs,* edited by J. S. Ames, Vol. IX, *The Laws of Gravitation.* See p. 136.)

NEWTON'S LAW OF ATTRACTION FOR A SPHERE.

The gravitational attraction of a particle toward a sphere, whose density may be a function of the distance from the center, is the same as if the mass of sphere were concentrated in a particle at the center of the sphere.

The attraction by such a sphere on a particle within it is due entirely to that portion of the sphere which lies inside of a concentric surface through the particle. In other words, the attractions due to all parts of the sphere which lie farther than the particle from the center cancel each other.

(Consult Thomson and Tait, *Treatise on Natural Philosophy,* Part II, *arts.* 462, 471 *et seq.*)

WEIGHT AND MASS.

Bodies are constant in mass but variable in weight, and weight is always proportional to mass.

Thus, $W = Mg$,

where W is the weight of the body, M its mass and g the acceleration of gravity.

The value of g increases from the equator to the poles. $g = 978.00$ cm. per sec. per sec. at the equator and $g = 983.01$ at the poles.

(Jeans, *Theoretical Mechanics*, pp. 29–31. For values of g at various places consult Chwolson, *Traité de Physique,* Vol. I, Part 3, p. 387. For accepted *mean* value of g consult Hering, *Conversion Tables,* p. 87.)

LAW OF FALLING BODIES.

The law by which bodies fall in vacuum is: The velocity of descent is proportional to the time of falling, and the distance of descent is proportional to the square of the time of falling.

Thus, $v \propto t$, and $s \propto t^2$.

Or, $v = gt$ and $s = \frac{gt^2}{2}$ when the body starts from rest.

Or, $\frac{d^2s}{dt^2} = g$, which integrated gives,

$$s = \tfrac{1}{2} gt^2 + A_1 t + A_2.$$

Here g is the constant acceleration of gravity. A_1 and A_2 are constants of integration.

(Consult Mach, *Science of Mechanics*, p. 130 *et seq*. See "Use of Analogy in Viewing Physical Phenomena" by E. F. Northrup, *Jour. Frank. Inst.*, p. 17, July, 1908.)

DESCENT ON AN INCLINED PLANE.

The acceleration along an inclined plane AC is to the acceleration along the vertical AB as the length AB is to the length AC, or,

$$\frac{\text{acceleration along inclined plane}}{\text{acceleration of gravity}} = \frac{AB}{AC} = \sin a, \text{ where } a$$

= the angle of inclination of the plane.

(Consult Mach, *Science of Mechanics*, pp. 137–138.)

CONDITION OF EQUILIBRIUM FOR A SYSTEM.

When a system of particles is in equilibrium under the action of any system of external forces, the sum of the components of all these forces in any direction is zero; and the sum of the moments of all these forces about any line is zero.

(Consult Jeans, *Theoretical Mechanics,* p. 64. For complete treatment, see Thomson and Tait, *Treatise on Natural Philosophy,* Part II, Chaps. VI and VII.)

BASIC EQUATIONS OF MECHANICS; COMMENT ON.

Let s = distance, t = time, v = instantaneous velocity, a = acceleration of a uniformly accelerated motion, F = ma = force and m = mass. Assuming the body starts from rest the relations of these quantities may be placed in the two groups;

$$\left.\begin{aligned} v &= at \\ s &= \tfrac{1}{2}at^2 \\ as &= \tfrac{1}{2}v^2 \end{aligned}\right\} \text{I} \qquad\qquad \left.\begin{aligned} mv &= Ft \\ ms &= \tfrac{1}{2}Ft^2 \\ Fs &= \tfrac{1}{2}mv^2 \end{aligned}\right\} \text{II}$$

Equations of group I contain the quantity a, and in addition two of the quantities s, t, v, as in table A, and equations of group II contain the quantities m, F, s, t, v, each equation containing m F and in addition to m F two of the quantities s, t, v, as in table B.

$$\text{(A)}\quad a \left\{\begin{array}{l} v, t \\ s, t \\ s, v \end{array}\right. \qquad\qquad \text{(B)}\quad m, F \left\{\begin{array}{l} v, \\ s, t \\ s, v \end{array}\right.$$

(Mach, *Science of Mechanics,* pp. 269–270.)

GENERAL MECHANICAL PRINCIPLE.

"A system always tends to move from rest in such a way as to diminish the potential energy as much as possible, and the force tending to assist a displacement in any direction is equal to the rate of diminution of the potential energy in that direction."

(J. J. Thomson, *Elements of Electricity and Magnetism*, p. 82.)

KEPLER'S FIRST LAW.

The planets move about the sun in ellipses, at one focus of which the sun is situated.

(Consult Mach, *Science of Mechanics*, p. 187.)

KEPLER'S SECOND LAW.

The radius vector joining each planet with the sun describes equal areas in equal times.

(This second law, the law of areas, can be explained simply if it be assumed, as a particular case, that the acceleration toward the sun is constant.)

(Consult Mach, *Science of Mechanics*, p. 188.)

KEPLER'S THIRD LAW.

The cubes of the mean distances of the planets from the sun are proportional to the squares of their times of revolution about the sun. This law may be stated mathematically,

$$\frac{R_1^3}{T_1^2} = \frac{R_2^3}{T_2^2} = \frac{R_3^3}{T_3^2} = \ldots\ldots = \text{a constant.}$$

Here R_1, R_2, R_3, are mean radii and T_1, T_2, T_3, etc., the respective times of revolution of the planets.

(Consult Mach, *Science of Mechanics*, pp. 188–189. Also Laplace, *Traité de Mécanique Céleste*, Book II, Chap. I.)

LAW OF THE LEVER.

Any lever is in equilibrium when the algebraic sum of the statical moments taken about the fulcrum equals zero.

By the statical moment is here meant the force acting at a point multiplied by the perpendicular distance from the line of support to the direction of the force.

(For a lucid discussion of the principle of the lever see Mach, *Science of Mechanics*, Chap. I.)

COMPOSITION BY PARALLELOGRAM RULE.

If a parallelogram A B C D is so drawn that two of its sides, as A B, A D, which meet in a point, represent as regards both magnitude and direction, two displacements, or two velocities, or two accelerations, or two forces to be compounded, then the diagonal A C of this parallelogram will represent, in respect to both magnitude and direction, the resultant displacement, velocity, acceleration or force.

(Consult Jeans, *Theoretical Mechanics*, Chaps. I and III.)

PARALLELOGRAM OF FORCES.

Two forces acting at the same time and for the same time upon a particle produce accelerations which are independent of each other and proportional to the forces. The resultant distance, direction, velocity or acceleration is obtained by compounding by the parallelogram rule.

(Consult Jeans, *Theoretical Mechanics*, Chap. III.)

LAMI'S THEOREM.

"When a particle is acted on by three forces, the necessary and sufficient condition for equilibrium is that the three forces shall be in one plane and that each force shall be proportional to the sine of the angle between the other two."

(For proof see Jeans, *Theoretical Mechanics*, p. 40.)

CENTRIPETAL ACCELERATION.

The centripetal acceleration of a body moving in a circle is directed toward the center of the circle and is equal to the square of its instantaneous linear velocity divided by its distance from the center about which it rotates at the instant considered.

For uniform rotation in a circle,

$$\text{acceleration} = \frac{\overline{\text{velocity}}^{\,2}}{\text{radius}}, \text{ or, } a = \frac{v^2}{r}.$$

(Consult Laplace, *Traité de Mécanique Céleste,* Book I, *art.* 10.)

PRINCIPLES OF THE SCREW AND THE WRENCH.

Any motion can be reduced to a translation and a rotation about an axis parallel to the translation. Such a combination of a translation and a rotation is called a Screw.

Similarly any system of forces acting on a rigid body may be replaced by a single force and a couple about the line of the force. This combination of a force and a couple is known as a Wrench.

Thus a screw and a wrench are the most general types of motion and force respectively.

(Jeans, *Theoretical Mechanics,* pp. 91, 106–107. For general treatment of motion in three dimensions, see Routh, *Elementary Rigid Dynamics,* Chap. V, pp. 184–228.)

THEOREM OF PRINCIPAL AXES OF INERTIA.

In every rigid body there are three mutually perpendicular axes of rotation intersecting at the center of gravity of the body which are characterized by the property that, if the body be rotated about any one of these axes, it will continue thus to rotate unless acted on by an external force. If rotating about any other axis the centrifugal forces tend to change the axis of rotation and the rotation is therefore unstable. The position of these axes is determined mathematically by the fact that, if they be taken as the axes of coördinates, the products of inertia.

$$\Sigma mxy, \ \Sigma myx, \ \Sigma mzx$$

for the body will vanish.

(Webster, *The Dynamics of Particles and of Rigid, Elastic and Fluid Bodies,* pp. 228–229. Also Routh, *Elementary Rigid Dynamics,* pp 12–13.)

MOMENT OF INERTIA AND ENERGY OF ROTATION.

The moment of inertia about any straight line through a rigid body made up of a system of particles is the sum of the products of the mass of each particle by the square of its perpendicular distance from this straight line; its axis of rotation.

Thus, moment of inertia $= I = \Sigma mr^2$ where m is the mass of a particle and r its perpendicular distance from its axis of rotation.

The kinetic energy E_r of the rotating body is, $E_r = \frac{1}{2} I \omega^2$, where ω is its angular velocity. This is the analogue of $E_1 = \frac{1}{2} mv^2$, where m is the mass of a body having a rectilinear velocity v.

(Thomson and Tait, *Treatise on Natural Philosophy,* Part I, *art.* 281.)

RADIUS OF GYRATION. (1)

The radius of gyration of any body (made up of a system of particles) about any axis is the perpendicular distance from that axis at which, if the whole mass were placed, it would have the same moment of inertia as before.

Mathematically the radius of gyration is,

$$k = \sqrt{\frac{\Sigma mr^2}{\Sigma m}}.$$

The kinetic energy of rotation of the body when its angular velocity is ω is:

$$E_r = \frac{1}{2}\omega^2 \Sigma mr^2 = \frac{1}{2}\omega^2 k^2 \Sigma m.$$

(Consult Thomson and Tait, *Treatise on Natural Philosophy,* Part I, *art.* 281.)

RADIUS OF GYRATION. (2)

If the radius of gyration of any body revolving about a line through its center of mass is k, then the radius of gyration K about any parallel line a distance a from this line is given by the relation,

$$K^2 = k^2 + a^2.$$

(Consult Jeans, *Theoretical Mechanics,* p. 291.)

ANALOGUES IN TRANSLATION AND ROTATION.

Force is measured by the product, mass × linear acceleration, or $F = ma$.

Moment of force is measured by the product, moment of inertia × angular acceleration, or $F_r = I\alpha$.

One-half the mass × $\overline{\text{linear velocity}}^2$ = *kinetic energy of translation,* or $E_t = \frac{1}{2} mv^2$.

One-half the moment of inertia × the $\overline{\text{angular velocity}}^2$ = *kinetic energy of rotation,* or $E_r = \frac{1}{2} I \omega^2$.

When no external forces are acting, the product, mass × velocity is constant, or mv = *linear momentum* = a constant.

Under similar conditions the product, moment of inertia × angular velocity is constant, or $I\omega$ = *angular momentum* = a constant.

(Ames, *Theory of Physics,* Chap. II.)

SIMPLE HARMONIC MOTION.

A body is said to oscillate with Simple Harmonic Motion when its acceleration is always directed toward the middle point of its path of oscillation and is proportional to its displacement therefrom.

The acceleration is given by the formula,

$$a = -4\pi^2 N^2 S,$$

where N is the frequency or number of complete oscillations in the unit of time and S is the displacement of the point from its middle position at any instant.

(Thomson and Tait, *Treatise on Natural Philosophy,* Part I, *art.* 57,)

SIMPLE PENDULUM.

The oscillation of a pendulum may be regarded as Simple Harmonic Motion when the amplitude of the oscillation is small. The approximate time of complete vibration is,

$$T = 2\pi\sqrt{\frac{l}{g}}.$$

Here l = the length of the simple pendulum and g = the acceleration of gravity.

(Jeans, *Theoretical Mechanics,* pp. 259–262. Also *Ganot's Physics, art.* 56.)

LAW OF THE COMPOUND PENDULUM.

If y is the length of a simple pendulum which oscillates in the same time as a compound pendulum, then the principle of the center of oscillation asserts that

$$y = \frac{\Sigma mr^2}{\Sigma mr}.$$

The compound pendulum executes isochronous oscillations when the amplitude of the oscillations is small. The time of a complete oscillation is that of a simple pendulum having this length y.

Here Σmr is the sum of the products of the elements of mass of the pendulum multiplied by their distances from the point of support and Σmr^2 is the moment of inertia of the pendulum about its axis of oscillations. Its complete period for oscillations of small amplitude is

$$T = 2\pi\sqrt{\frac{\Sigma mr^2}{g\,\Sigma mr}}.$$

(Mach, *Science of Mechanics,* pp. 173–177.)

CONVERTIBILITY OF CENTER OF OSCILLATION AND POINT OF SUSPENSION.

When the center of oscillation and the point of suspension of a compound pendulum are interchanged the time of oscillation remains the same.

This principle was employed by Captain Kater for determining the exact length of the seconds pendulum. Hence the term "Kater's Pendulum."

(Mach, *Science of Mechanics,* p. 172 *et seq.* and p. 186. Also *Ganot's Physics, art.* 81.)

CYCLOIDAL PENDULUM.

A particle subjected to the force of gravity and constrained to move in a cycloidal path will have a harmonic motion which is strictly isochronous whatever its amplitude of oscillation.

The period will be,

$$T=2\pi\sqrt{\frac{2D}{g}},$$

where D is the diameter of the rolling circle which generates the cycloid. Thus the period is that of a simple pendulum of length 2D.

(Jeans, *Theoretical Mechanics,* p. 267.)

MINIMUM POTENTIAL ENERGY.

When a system is in stable equilibrium its potential energy is as small as possible; that is, any small movement imparted to the system, subject to its degrees of freedom, will increase and not decrease its potential energy.

(For a mathematical discussion see Jeans, *Theoretical Mechanics,* p. 174 *et seq.*)

IMPACT BETWEEN TWO BODIES.

When two bodies meet in impact the *impulse of restitution* I' equals the *impulse of compression* I times a certain coefficient e called the coefficient of elasticity or coefficient of resilience.

Thus, $I' = eI$.

The coefficient e is unity for perfectly elastic bodies and zero for perfectly inelastic bodies.

Its value for actual bodies must be obtained by experiment.

(Jeans, *Theoretical Mechanics*, pp. 238–241.)

NEWTON'S EXPERIMENTAL LAW OF IMPACT.

When two bodies meet in impact and their centers of gravity lie, at the moment of impact, in a line through the point of contact, then the normal component of relative velocity of their centers of gravity after impact is equal to the relative velocity before impact times the coefficient of resilience, and is in the opposite direction, or

$$v - v' = -e\,(u - u')$$

where v, v' are the velocities of the two bodies after impact, u, u' their velocities before impact and e the coefficient of resilience.

(Jeans, *Theoretical Mechanics*, pp. 244, 245.)

RELATION CONNECTING VELOCITIES OF TWO BODIES BEFORE AND AFTER IMPACT.

If two bodies of masses m and m' have velocities u, u' before impact and velocities v, v' after impact, then by the conservation of momentum

$$mv + m'v' = mu + m'u',$$

or, if U is the common velocity of the two bodies after impact,

$$U = \frac{mu + m'u'}{m + m'} \quad \text{(called the rule of Wallis).}$$

(Jeans, *Theoretical Mechanics*, p. 245. Also Mach, *Science of Mechanics*, p. 318.)

RELATION OF VELOCITIES AFTER AND BEFORE IMPACT.

The normal velocities v, v′ of two impacting bodies of masses m, m′ after collision are given in terms of their velocities u, u′ before collision by the two relations,

$$v = \frac{mu + m'u' - em'(u - u')}{m + m'},$$

$$v' = \frac{mu + m'u' + em(u - u')}{m + m'},$$

where e is the coefficient of resilience.

(Jeans, *Theoretical Mechanics*, p. 245.)

IMPACT.

Relative velocity of approach and recession of impinging *perfectly elastic* bodies is the same.

Also, $m'v' + mv = m'u' + mu$

and, $m'v'^2 + mv^2 = m'u'^2 + mu^2$,

namely, the quantity of motion before and after impact, estimated in the same direction is the same, also, the *vis viva,* or kinetic energy of the system.

(Consult Mach, *Science of Mechanics,* p. 322. See Thomson and Tait, *Treatise on Natural Philosophy,* Part I, *arts.* 300–301, for comments upon conditions which determine loss of kinetic energy with impact.)

TIME OF ELECTRICAL CONTACT OF IMPACTING STEEL-SPHERES.

When two equal steel-spheres come together the time in millionths of a second during which they are in contact equals 74.7 times the diameter (in cm.) divided by the fifth root of their velocity of approach.

Thus, $T_\mu = 74.7 \frac{D}{v^{1/5}}$

where T_μ = microseconds,

D = diameter of spheres in centimeters,

v = velocity in centimeters per second with

which they approach each other at the moment of the beginning of contact.

(This law was experimentally determined by E. F. Northrup and A. E. Kennelly. See *Jour. Frank. Inst.*, July, 1911.)

VELOCITY OF A DISTURBANCE IN AN ELASTIC MEDIUM.

The velocity with which any compressional disturbance is propagated through an elastic medium is given by the equation

$$V = \sqrt{\frac{E}{D}}$$

where E is the coefficient of volume elasticity and D is the density of the medium. This law is due to Newton.

(Watson, *A Text Book of Physics*, pp. 362–365.)

BERTRAND'S PRINCIPLE OF SIMILITUDE.

If two systems differ only in geometrical magnitude, the masses of corresponding parts being proportional to each other, they will be mechanically similar if the forces acting on and the velocities of respective parts of the systems bear the relation

$$\text{force proportional to } \frac{\text{mass} \times \overline{\text{velocity}}^{2}}{\text{linear dimensions}}.$$

This principle gives the conditions under which two systems which are geometrically similar may also be mechanically similar and is important in testing small model machines.

(Routh, *Elementary Rigid Dynamics*, pp. 234–297. For a full discussion, see article by Bertrand in *Cahier 32*, of the *Journal de l'Ecole Polytechnique.*)

CHANGE OF MOMENTUM DUE TO AN IMPULSIVE FORCE.

When a body moves in such a way that its configuration with respect to the force which acts upon it remains always the same, the moving force is measured by the rate of increase of the momentum.

If F is the force, p the momentum and t the time,

$$F = \frac{dp}{dt} \quad \text{whence}$$

$$p = \int F dt.$$

This time-integral is called the *Impulse of the Force.*

(Maxwell, *Treatise on Electricity and Magnetism,* Vol. II, *art.* 558.)

SLIDING FRICTION.

The tangential force required to slide one body over another is independent of the extent of their surfaces in contact and of the rate of sliding. It is proportional to the normal force R pressing them together and to a constant for the two bodies called the coefficient of friction μ. Thus $F = \mu R$.

Distinction is made between *statical friction* where a relative motion of the bodies is just not produced by the acting forces and *kinetic friction* where there is relative motion of the bodies in contact. Kinetic friction is generally less than the extreme force of static friction.

(Thomson and Tait, *Treatise on Natural Philosophy,* Part II, *arts.* 450–452.)

ROLLING FRICTION.

Rolling friction acts much as does sliding friction. The coefficient u is in this case much smaller and the force required to cause motion is again proportional to the normal force R pressing the bodies together.

Thus $F = uR$.

(*Ganot's Physics, art.* 49.)

HOOKE'S LAW.

Hooke's law states that within the elastic limit of any body the ratio of the stress acting upon the body to the strain produced is constant. This constant ratio of stress to strain for any particular type of change in any body is called its "coefficient of elasticity."

(Ames, *Theory of Physics,* pp. 103–104. Also see *Ganot's Physics, art.* 85.)

YOUNG'S MODULUS.

By "Young's Modulus" is understood the force which would be required to stretch a body of unit cross-section to double its length if such lay within the elastic limit. It is a constant for any one material.

Thus, Young's Modulus $= M = \frac{LF}{ea}$,

where L = length of body (usually a wire),

a = area of cross-section,

F = stretching force applied and

e = total elongation produced by F.

Similar moduli hold for compression, flexure and torsion.

(Crew, *General Physics,* p. 131. Also *Ganot's Physics, art.* 87. Also Thomson and Tait, *Treatise on Natural Philosophy,* Part II, *arts.* 686–688.)

CONSERVATIVE SYSTEMS.

A system is called a conservative system when it is such that the total work done in performing any series of displacements which bring the system back to its original configuration is algebraically zero.

(Consult Thomson and Tait, *Treatise on Natural Philosophy,* Part I, *art.* 271 *et seq.*)

PRINCIPLE OF WORK DONE BY A CONSERVATIVE SYSTEM.

When a conservative system is changed from one configuration to another the work done is independent of the manner in which the change is made and depends only on the initial and final states of the system.

(Jeans, *Theoretical Mechanics,* p. 164.)

RAYLEIGH'S RECIPROCATION THEOREM.

If a system of bodies is struck successively at two different points a and b by impulsive forces P_1 and Q_2, each blow will in general affect all the bodies of the system. Let u_1, u_2 be the velocities of the points a and b produced by the blow P_1 and v_1, v_2 be the velocities produced by the blow Q_2. Rayleigh's Theorem states that

$$P_1 v_1 = Q_2 u_2.$$

(Routh, *Elementary Rigid Dynamics,* pp. 338, 339. Many illustrations are given in Rayleigh's *Theory of Sound.*)

PERPETUAL MOTION IMPOSSIBLE.

The impossibility of perpetual motion rests upon the following consideration: Let a system be made to pass by frictionless constraint from a configuration A to another configuration B and return by another path to A. Assume more work is done upon the system by the mutual forces when it follows one path than when it follows the other. Let this process be repeated over and over again forever. The system will then be a continual source of energy without the consumption of materials, something which all experience shows to be impossible.

(Consult Thomson and Tait, *Treatise on Natural Philosophy,* Part I, *art.* 272.)

CONSERVATION OF THE MOVEMENT OF THE CENTER OF GRAVITY.

The state of rest or movement of the center of gravity of several bodies is not altered by the reciprocal action of these bodies. Or, the motion of the center of gravity of a system of bodies is changed only by forces external to the system.

(For extension of principle consult Laplace, *Mécanique Céleste,* Vol. I, and Lagrange, *Mécanique Analytique,* Vol. I, Part II, Section 1, *art.* 15, and Section 3, *art.* 1 *et seq.*)

PRINCIPLE OF THE MOTION OF THE CENTER OF GRAVITY.

The center of gravity of any system moves as if all the external forces were acting on the entire mass of the system concentrated in a particle at the center of gravity and all the external forces were applied to this particle.

(Jeans, *Theoretical Mechanics,* pp. 224–226.)

CONSERVATION OF MATTER OR MASS.

If a mass or quantity of matter of any kind be selected for consideration, then the "principle of the conservation of mass" asserts; that no changes of any kind which can occur in the mass, whether brought about by the action of forces internal or external to the mass, can alter in the slightest degree its total quantity.

For the measure of the total quantity of mass considered, may be taken, either its inertia or the gravitational attraction of the earth upon it when located at a particular place.

(Consult Ames, *Theory of Physics,* pp. 6, 7. Also see definition under word CONSERVATION, *New Century Dictionary.*)

LAW OF CONSERVATION OF MOMENTUM. (1)

When any system of particles moves without being acted upon by any external force, the total momentum of the system remains constant in magnitude and direction, and the moment of momentum about any axis remains constant.

(Consult Mach, *Science of Mechanics,* p. 288.)

LAW OF CONSERVATION OF MOMENTUM. (2)

If all velocities *in a given direction* are reckoned as *positive* and all in the opposite direction as *negative,* then the sum of the momenta of a system of bodies, uninfluenced by forces external to the system, is preserved constant, whether or not these bodies meet in impact and whether or not they are elastic or inelastic.

(For mathematical definition, $\Sigma m \frac{dx}{dt} =$ a constant, see Appell, *Traité de Mécanique Rationnelle,* Vol. II, p. 20. Consult Mach, *Science of Mechanics,* p. 326.)

LAW OF THE CONSERVATION OF MOMENT OF MOMENTUM.

Let a system of particles of masses m_1, m_2, m_3, etc., have velocities u_1, u_2, u_3, etc., in a plane or in planes which are parallel. If O be the position of *any* fixed line perpendicular to these planes and r_1, r_2, r_3, etc., the perpendicular distances from this line to the directions in which the various particles are moving at a given instant the products, m u r, are "moments of momenta" and the law of the conservation of moment of momentum states that the $\Sigma mur =$ a constant, if no forces external to the system act upon it. In taking the sum of the products, the products are reckoned positive or negative according as a particle viewed from O is moving toward the right or toward the left.

(Ames, *Theory of Physics,* p. 49. Consult also Lagrange, *Mécanique Analytique,* Vol. I., Part II, Section 1, *art.* 16, p. 260.)

CONSERVATION OF LIVING FORCES.

This principle states that the difference of the force-functions (or work) at the beginning and at the end of the motion of a system is equal to the difference of the *vires vivæ* (kinetic energies) at the beginning and the end of the motion. Namely,

$$\Sigma(U - U_0) = \Sigma \tfrac{1}{2} m (v^2 - v_0^2).$$

(For a formal development of this principle see Mach, *Science of Mechanics*, p. 478 *et seq.*)

PRINCIPLE OF THE CONSERVATION OF AREAS (D'ARCY'S STATEMENT).

According to d'Arcy, the principle states that the sum of the products of the mass of each body by the area which its radius vector describes about a fixed center on any plane of projection is always proportional to the time. Or,

$$\Sigma mA = Kt \quad \text{or} \quad \Sigma m \frac{x\,dy - y\,dx}{2} = K\,dt.$$

(See Laplace, *Mécanique Céleste*, Vol. I, Part I, Book I, *art.* 21.)

PRINCIPLE OF THE CONSERVATION OF AREAS (MACH'S STATEMENT).

Mach states this principle as follows: If from any point in space radii be drawn to several masses and projections be made upon any plane of the areas which the several radii describe, the sum of the products of these areas into the respective masses will be independent of the action of internal forces.

(Mach, *Science of Mechanics*, p. 294.)

PRINCIPLE OF VIRTUAL VELOCITIES.

If the point of application of a force be displaced through a small space, the resolved part of the displacement in the direction of the force has been called its "Virtual Velocity." Mathematically expressed the principle asserts that for a body to be in equilibrium

$$Pp + P'p' + P''p'' + \ldots\ldots = 0.$$

Here P, P′, P″, etc., are applied forces acting upon a connected system at points A, B, C, etc., and p, p′, p″, etc., are the projections on the lines of the forces of small displacements of the points. The projections are to be taken positive when they fall in the direction of a force and negative when they fall in the opposite direction.

(See Mach, *Science of Mechanics*, Chap. I, Sec. IV, p. 49 *et seq.* for a clear and full discussion of this principle.)

PRINCIPLE OF LEAST ACTION.

In the movement of bodies which interact so that the total energy remains constant, the sum of the products of the masses by the velocities and by the spaces described is a minimum. This principle was extended to systems of masses by Lagrange who presented it in the form,

$$\delta \Sigma m \int v ds = 0.$$

(See Mach, *Science of Mechanics*, pp. 364–380 for a good physical discussion of this rather obscure principle. It is mathematically treated in Thomson and Tait, *Treatise on Natural Philosophy*, Part 1, *arts.* 326, 327.)

KELVIN'S MINIMUM-ENERGY THEOREM.

If a material system, initially at rest, is set in motion by impulses applied to particular points in such a way that these points acquire specified velocities, the motion of the entire system is such as to make the total kinetic energy of the system less than it would be in any other possible motion of the system consistent with the same velocity conditions.

(Houstoun, *An Introduction to Mathematical Physics*, p. 67.)

STOKES' LAW FOR THE FALL OF A SPHERE THROUGH A VISCOUS MEDIUM.

When a small sphere falls under the action of gravity through a viscous medium it ultimately acquires a constant velocity equal to

$$v = \frac{\frac{2}{9} g a^2 (d_1 - d_2)}{k},$$

where a is the radius and d_1 the density of the sphere, d_2 is the density of the medium and k is its coefficient of viscosity.

When great accuracy is required, correction factors must be added to the above expression.

This formula has been of much service in determining the charge on an electron.

(G. G. Stokes, *Mathematical and Physical Papers*, Vol. III, p. 59. See also Campbell, *Modern Electrical Theory*, p. 91.)

D'ALEMBERT'S PRINCIPLE. (1)

When forces act upon one or more rigidly connected points of a system of masses, these forces, called the *impressed* forces, may each be resolved into two components, the *equilibrated* forces and the *effective* forces. The latter only are operative in producing motion, while the former form a system balanced by the connections. The sum of the products of the effective forces by the elementary displacements which they produce is equal to the element of work performed upon the system.

(Mach, *Science of Mechanics*, pp. 335–343.)

D'ALEMBERT'S PRINCIPLE. (2)

All of the work performed on any system is performed by the effective components of the impressed forces. When no work is performed the system is in equilibrium. The two mathematical forms in which D'Alembert's principle is usually expressed are:

$$\Sigma[(X - ma)\,\delta x + (Y - mb)\,\delta y + (Z - mc)\,\delta z] = O,$$

$$\Sigma(X\delta x + Y\delta y + Z\delta z) = \Sigma m\,(a\delta x + b\delta y + c\delta z).$$

Here X, Y, Z are the mutually perpendicular components parallel to rectangular coördinates of every force P impressed on the masses m. ma, mb, mc are the corresponding components of every effective force W, where a, b, c denote accelerations and δx, δy, δz are displacements in the directions of the coördinates.

(Consult Mach, *Science of Mechanics*, p. 342.)

GAUSS'S PRINCIPLE OF LEAST CONSTRAINT. (1)

Let the masses M, M_1, etc., be joined in any manner with one another. If these masses were *free* they would describe in the element of time under the action of forces impressed on them, paths ab, a_1b_1, etc.; but in consequence of the connections, they describe in the same element of time the paths ac, a_1c_1, etc. Gauss's principle asserts that the motion of the connected points is such that, *for the motion actually taken,* the sum of the products of the mass of each particle into the square of the distance of its deviation from the position it would have reached if free, is a *minimum.*

(Consult Mach, *Science of Mechanics*, p. 350 *et seq.* for a clear development of this principle.)

GAUSS'S PRINCIPLE OF LEAST CONSTRAINT. (2)

The motion of a system of material points interconnected in any way and submitted to any influences, accords at each instant as closely as possible with the motion the points would have if they were free. The actual motion takes place so that the constraints on the system are the least possible. For the measurement of the constraint, during any element of time, is to be taken the sum of the products of the mass of each point by the square of its deviation from the position it would have occupied at the end of the element of time, if it had been free.

PRINCIPLE OF LEAST CONSTRAINT; COMMENT ON.

Gauss's principle of "least constraint" gives equations which when differentiated yield D'Alembert's principle. The kernel idea is "that the work of the forces which deviate the movement of the system from the paths it would take if unconstrained is as small as possible under the conditions." Gauss's principle includes both statical and dynamical cases.

(Clear expositions and analytical treatments of the principle of least constraint are to be found in Appel, *Traité de Mécanique Rationnelle,* and in Mach, *Science of Mechanics.*)

HAMILTON'S PRINCIPLE.

This principle is expressed as follows:

$$\delta\int_{t_0}^{t_1} (U + T)\,dt = 0, \text{ or } \int_{t_0}^{t_1} (\delta U + \delta T)\,dt = 0,$$

where δU and δT denote the variations of the work and *vis viva* vanishing for the initial and terminal epochs. "Hamilton's principle is easily deduced from D'Alembert's, and, conversely, D'Alembert's from Hamilton's." The principles, least action, least constraint, D'Alembert's, Gauss's and Hamilton's, are not expressions of different facts, but rather, are simply views of different *aspects* of the same fact.

(Mach, *Science of Mechanics*, pp. 380–384. Also consult Appell, *Traité de Mécanique Rationnelle,* Vol. II, p. 422.)

PRINCIPLE OF THE CONSERVATION OF ENERGY.

In every modification of a material system, not affected by forces foreign to the system, the sum of its potential and kinetic energies remains constant.

Calling E the kinetic energy and P the potential energy of the system, $E + P = K$, a constant.

(See Helmholtz's lecture delivered at Carlsruhe about 1862, "On the Conservation of Force," in *Popular Scientific Lectures,* Vol. I. Also consult Thomson and Tait, *Treatise on Natural Philosophy,* Part I, *arts.* 269–278.)

DIFFERENT STATES OF EQUILIBRIUM.

A body is in stable equilibrium when a slight movement from its position will raise its center of gravity. It is in unstable equilibrium when such movement will lower its center of gravity. It is in neutral equilibrium when such movement will neither raise nor lower its center of gravity.

(*Ganot's Physics, art.* 72.)

II

HYDROSTATICS, HYDRODYNAMICS AND CAPILLARITY

HYDROSTATICS, HYDRODYNAMICS AND CAPILLARITY

ARCHIMEDES' PRINCIPLE.

When a body is in equilibrium in a fluid the fluid exerts an upward force on the body equal to the weight of the displaced fluid and acts through its center of gravity. Or a body immersed in liquid loses weight equal to the weight of the displaced liquid.

(This principle furnishes the most convenient method for the determination of the specific gravity of a body, in terms of that of the fluid used.)

(*Ganot's Physics, art.* 112.)

EQUILIBRIUM OF FLOATING BODIES.

1. The floating body must displace a volume of liquid whose weight equals that of the body.
2. The center of gravity of the floating body must be in the same vertical line with that of the fluid displaced.
3. The equilibrium of a floating body is stable or unstable according as the metacenter is above or below the center of gravity.

(*Ganot's Physics, art.* 114.)

PRESSURE PRODUCED IN LIQUIDS BY GRAVITY.

The pressure in each layer is proportional to the depth.

With different liquids and the same depth, the pressure is proportional to the density of the liquid.

The pressure is the same at all points of the same horizontal layer.

(*Ganot's Physics, art.* 98.)

HYDROSTATIC PARADOX.

The total weight or downward *force* exerted by a vessel containing liquid depends on the shape and size of the containing vessel and may be greater or smaller than the force which is applied to give the liquid its hydrostatic pressure.

(Consult *Ganot's Physics, art.* 102. Also Ames, *Theory of Physics,* pp. 112–114.)

PASCAL'S LAW.

The fluid pressure due to the reaction of the walls of the containing vessel is the same at all points throughout the fluid. Or pressure exerted anywhere upon a mass of liquid is transmitted undiminished in all directions, and acts with the same force on all equal surfaces and in a direction at right angles to those surfaces.

(Kimball, *College Physics,* p. 111. Also *Ganot's Physics, art.* 97.)

CONDITION OF THE EQUILIBRIUM OF LIQUIDS.

1. Its surface must be everywhere perpendicular to the resultant of the forces which act on the molecules of the liquid.
2. Every molecule of the mass of the liquid must be subject in every direction to equal and contrary forces.

(Consult *Ganot's Physics, art.* 103. For Clairaut's mathematical statement of the general condition of liquid equilibrium, see Mach, *Science of Mechanics,* p. 397.)

EQUILIBRIUM OF LIQUIDS IN COMMUNICATING VESSELS.

When two non-miscible liquids of different densities are placed in communicating vessels their free surfaces will stand at different heights above the surface of contact of the two liquids. Neglecting any capillary action, the heights of the free surfaces of the two liquids above their surface of contact are in inverse ratio to their densities.

(Ames, *Theory of Physics*, p. 115. Also Chwolson, *Traité de Physique*, Vol. I, Part 5, p. 569.)

RESULTANT OF FORCES OF COHESION AT THE SURFACE OF A LIQUID.

At the surface of a liquid all the forces of cohesion have a resultant which is directed toward the interior of the liquid normally to its surface.

(Chwolson, *Traité de Physique*, Vol. I, Part 5, p. 563.)

EFFLUX: TORRICELLI'S THEOREM.

The velocity of efflux of a stream of liquid issuing from a cistern is the velocity which a freely falling body would have on reaching the orifice after having started to fall from rest at the surface-level of the fluid. (Only strictly true when friction and shape of the orifice are disregarded.) The velocity is,

$$v = \sqrt{2gh}$$

where g = acceleration of gravity, and
h = height of level of liquid above orifice.

(*Ganot's Physics, art.* 142. Also Mach, *Science of Mechanics*, p. 402.)

QUANTITY OF EFFLUX: "VENA CONTRACTA."

The stream of water issuing from a cistern through a circular, sharp-edged orifice of area A contracts after leaving the orifice. The theoretical discharge would be $E = A\sqrt{2gh}$, but the actual discharge is found to be about $0.62\ A\sqrt{2gh}$, where g equals acceleration of gravity and h equals height of level of liquid above orifice.

The reduced velocity is due to the contraction of the cross-section of the stream, called the *vena contracta*.

(Consult *Ganot's Physics, art.* 145. Also Chwolson, *Traité de Physique*, Vol. I, Part 5, pp. 697, 698.)

BERNOULLI'S THEOREM: FLOW OF LIQUIDS.

At any point in a tube, through which a liquid is flowing, the pressure plus the potential energy due to position plus the kinetic energy remains constant (friction being disregarded). Or, along a horizontal stream-line, the relation holds,

$$\tfrac{1}{2}\ \text{density} \times \overline{\text{velocity}}^2 + \text{pressure} = \text{a constant}.$$

The general mathematical statement of the theorem is,

$$p + g\rho h + \tfrac{1}{2}\rho v^2 = c, \text{ a constant,}$$

where ρ = density of liquid,

g = acceleration of gravity,

v = velocity of flow,

h = distance above any horizontal plane of reference to the point in the liquid considered and

p = the hydrostatic pressure.

(Consult Crew, *General Physics*, p. 149. Also Mach, *Science of Mechanics*, p. 413 *et seq.* Also Chwolson, *Traité de Physique*, Vol. I, Part 5, p. 690.)

FLOW OF WATER IN PIPES.

1. The loss of head from friction is proportional to the length of pipe.
2. It increases with the roughness of the interior surface.
3. It decreases as the diameter of the pipe increases.
4. It increases nearly as the square of the velocity.
5. It is independent of the pressure of the water.

(Merriman, *Treatise on Hydraulics*, p. 209.)

PRINCIPLE OF CONTINUITY.

In a state of steady flow, the quantity of fluid passing any cross-section of the stream in a given time is the same for all sections of the stream. Thus the product SV is constant, where S is the cross-section and V the velocity of the stream at any point.

The "Equation of Continuity," which expresses this fact mathematically is,

$$\frac{1}{\delta}\frac{d\delta}{dt}+\frac{du}{dx}+\frac{dv}{dy}+\frac{dw}{dz}=0,$$

where u, v, w are the velocity-components in the directions of the x, y, z axes respectively at any point of the fluid, δ is the density of the fluid and t represents time.

If the fluid is incompressible the first term of this equation vanishes.

(Webster, *The Dynamics of Particles and of Rigid, Elastic and Fluid Bodies*, pp. 496–499.)

RESISTANCE TO THE MOTION OF A SOLID THROUGH A FLUID.

The pressure against a solid moving through a fluid is approximately proportional to the square of their relative velocity.

(Values of the constant of proportionality for the important case of wind against a sail or aeroplane are given in *Smithsonian Physical Tables*, p. 124.)

VORTEX MOTION.

Definitions: A Vortex Line is a curve whose tangent at every point coincides with the direction of the instantaneous axis of rotation at that point. A space bounded by vortex lines is called a Vortex Tube and the enclosed fluid is said to have a Vortex Motion. The strength of such a tube at any cross-section normal to its axis is defined as the product of the angular velocity w and the cross-sectional area S of the tube at that point.

Laws:

1. A vortex tube always contains the same elements of fluid.
2. The strength wS of a vortex tube is the same at all parts of the tube and does not change with time (in a perfect fluid.)
3. Vortex tubes are either closed surfaces or have their extremities in the surface of the fluid.

(Webster, *The Dynamics of Particles and of Rigid, Elastic and Fluid Bodies,* pp. 509–511. For a complete treatment see Helmholtz, *Uber Integrale der hydrodynamischen Gleichungen, welche den Wirbelbewegungen entsprechen, Wissenschaftliche Abhandlungen,* Vol. I, p. 101. For a photographic study of vortex motions in water see "An Experimental Study of Vortex Motions in Liquids," by E. F. Northrup, *Jour. of the Franklin Institute,* Sept. and Oct., 1911.)

FLOW THROUGH CAPILLARY TUBES. POISEUILLE'S LAW.

The volume of liquid V which will flow in unit time through a capillary tube of length l and radius r, is given by the formula $v = \frac{\pi p r^4}{8 k l}$. In C.G.S. units, p = pressure-difference over the length l of the tube in dynes per square centimeter and k is the coefficient of internal friction or viscosity.

The reciprocal of the viscosity, namely $\frac{1}{k}$, is called the fluidity.

This law was discovered and investigated by M. Poiseuille in 1843.

(Consult Poynting and Thomson, *Properties of Matter,* pp. 207–209. Also Chwolson, *Traité de Physique,* Vol. I, Part 5, p. 674.)

HYDRODYNAMICAL THEOREM.

"If the bounding surface of a liquid, primitively at rest, be made to vary in a given arbitrary manner, the *vis viva* of the entire liquid at each instant will be less than it would be if the liquid had any other motion consistent with the given motion of the bounding surface."

(*Mathematical and Physical Papers,* by Lord Kelvin, Vol. I, p. 109.)

CAPILLARY ACTION: JURIN'S LAW.

For the same liquid and the same temperature, the mean height of the ascent in a capillary glass-tube is inversely as the diameter of the tube. Thus diameter × height = a constant.

(*Ganot's Physics, art.* 131. Also consult Poynting and Thomson, *Properties of Matter,* p. 140. Also, Chwolson, *Traité de Physique,* Vol. I, Part 5, p. 616.)

LAW OF CAPILLARY ACTION. (1)

For various liquids and the same temperature and in tubes of the same diameter the mean heights to which the liquids will rise vary with the nature of the liquid. (Of all liquids water rises the highest.)

(*Ganot's Physics, art.* 131.)

LAW OF CAPILLARY ACTION. (2)

For the same liquid and the same temperature, the mean heights to which the liquid rises are independent of the form of the capillary tube except at the meniscus. Provided the liquid moistens the tube, neither the thickness of the tube nor its nature has any influence on the height to which the liquid rises.

(*Ganot's Physics, art.* 131.)

LAW OF CAPILLARY ACTION. (3)

The height to which a liquid rises in a capillary tube diminishes as the temperature increases. As the height becomes less the meniscus becomes flattened. When the sides of the tubes are not moistened Jurin's law holds approximately for the depression of the liquid.

(*Ganot's Physics, art.* 131.)

CAPILLARY CORRECTIONS OF MERCURY COLUMNS.

"The height of the meniscus and the value of the capillary depression depend on the bore of the tubing, on the cleanliness of the mercury and on the state of the walls of the tube. The correction is negligible for tubes with diameters greater than about 2.5 cms."

(See Kaye and Laby, *Physical and Chemical Constants,* p. 17. Or see *Smithsonian Physical Tables,* p. 123.)

SURFACE TENSION AND WORK OF THE FORCES OF COHESION.

Every diminution in the extension of a liquid-surface is associated with work done by forces of molecular cohesion. Every augmentation of a liquid-surface is associated with work done by exterior forces; the result of which is an augmentation of the store of potential energy in the liquid and the quantity of this potential energy depends upon the area of the liquid-surface.

(Chwolson, *Traité de Physique,* Vol. I, Part 5, p. 590.)

NORMAL PRESSURE ON A LIQUID-SURFACE.

The forces of cohesion which act upon the molecules of the superficial layer of a liquid result in producing a certain normal pressure per unit area upon the surface of the liquid. The magnitude of this pressure depends upon the form of the surface. If K is its magnitude for a plane surface, the normal pressure is greater than K on a convex surface and less than K on a concave surface.

(Chwolson, *Traité de Physique,* Vol. I, Part 5, p. 590.)

FORM ASSUMED BY A LIQUID-MASS UNDER THE INFLUENCE OF SURFACE TENSION ALONE.

A liquid-mass not subjected to any exterior force is in equilibrium. The pressure exerted upon it by forces resulting from surface-tension is the same at all points of the surface of the mass and the surface of the liquid assumes at all points the same mean curvature.

(Chwolson, *Traité de Physique,* Vol. I, Part 5, p. 601.)

PRESSURE-DIFFERENCE ON THE TWO SIDES OF A SOAP-FILM.

The pressure-difference p on the two sides of a soap-film is given by the relation,

$$p = 2T\left(\frac{1}{R_1} + \frac{1}{R_2}\right).$$

Here T is the surface-tension of the film, and R_1, R_2 are the two radii of Principal Curvature of the surface of the liquid-film at the point considered. The pressure on the concave side always exceeds that on the convex side of the film. In the formula above the convention must be made that a radius of Principal Curvature is to be taken positive or negative according as the corresponding center of curvature falls on the side of the surface where the pressure is greater or on the opposite side. For the case of a spherical soap-film $R_1 = R_2$ and thus the normally acting inward pressure on the inside of the film exceeds that on the outside by an amount

$$p = \frac{T}{R}$$

where R is the radius of the spherical film.

(Poynting and Thomson, *Properties of Matter*, pp. 144–146. Also consult Chwolson, *Traité de Physique,* Vol. I, Part 5, Chap. IV.)

VERTICAL DISTANCE BETWEEN TWO ELEMENTS OF A LIQUID-SURFACE.

If two elements of the surface of an extended liquid-mass resting on a plane surface be selected such that one of these elements is horizontal and the other vertical, then the perpendicular distance between the two elements is equal to the square root of the capillary constant of the liquid. Or,

$$h = \sqrt{k} = \sqrt{\frac{2T}{\delta}}$$

where k is the capillary constant, T the surface-tension and δ the density of the liquid.

(Chwolson, *Traité de Physique,* Vol. I, Part 5, pp. 628–630.)

ACTION ON LIGHT FLOATING BODIES OF SURFACE-TENSION.

Surface-tension will cause an attraction between two floating bodies both of which are wet if the curved portions of the liquid surrounding each of them intersect; the same is true if both the bodies are not wet. If one is wet and the other is not they will repel each other if the two curved portions of the liquid intersect.

(For an account of various capillary phenomena see *Ganot's Physics, art.* 136.)

WATER-WAVES, SPEED OF.

1. In water where the depth h is small compared with the wave-length, the speed of propagation V of the wave-crest is,

$$v = \sqrt{gh},$$

where g = the acceleration of gravity.

2. In deep water where the depth is large compared with the wave-length l, the speed is,

$$v = \sqrt{\frac{gl}{2\pi}}.$$

Thus a deep sea-wave 378 feet long travels with a speed of 44 feet per sec.

(Crew, *General Physics,* p. 188. Also *Mathematical and Physical Papers,* by Lord Kelvin, Vol. III, p. 519.)

RIPPLES ON SURFACE OF LIQUIDS.

Very short waves (less than 1.6 cm.), or ripples, may be considered to be propagated by the surface-tension of the liquid alone. Ripples of short wave-length travel faster than ripples of long wave-length which is just the reverse of what happens with ordinary water-waves propagated by gravity.

(For a general discussion of waves and ripples, consult Chwolson, *Traité de Physique,* Vol. I, Part 5, pp. 706–709.)

RIPPLES, SPEED OF.

1. Neglecting the effect of gravity upon the speed of short waves or ripples, the velocity is,

$$v = \sqrt{\frac{2\pi T}{l\rho}}.$$

where T is the surface-tension of the liquid, ρ its density and l is the wave-length.

2. When account is also taken of the acceleration g of gravity

$$v = \sqrt{\frac{2\pi T}{l\rho} + \frac{gl}{2\pi}}.$$

(The slowest water-waves are found to have a length of about 1.6 cm. Waves shorter than this are called "ripples.")

(Crew, *General Physics*, pp. 191, 192.)

III

SOUND

SOUND

DEFINITIONS.

1. *Sound* in the physical sense is either the vibrations of its source, or, of the elastic medium surrounding the source.
2. *Noise* is a sound resulting from irregular and practically unanalyzable vibrations.
3. *Musical* tones are distinguished by:
   a. their *force* determined by the amplitude of the vibrations,
   b their *pitch* determined by the frequency of the vibrations,
   c. their *quality* determined by the harmonics present.
4. A *periodic motion* is one which constantly returns to the same condition after equal intervals of time.

PROPAGATION OF SOUND.

Sound is propagated as a longitudinal wave in any elastic medium. It cannot be transmitted through vacuous space, as the presence of an elastic medium is essential.

(*Ganot's Physics, arts.* 224, 225.)

INTENSITY OF SOUND. (I)

The intensity of sound is inversely as the square of the distance of the sounding body from the ear. That is, sound radiates from a point in a homogeneous medium so that the wave-front is spherical in form.

(*Ganot's Physics, art.* 227.)

INTENSITY OF SOUND. (2)

The intensity of sound in a physical sense is the quantity of energy which traverses in the unit of time the unit of area normal to the sonorous ray. In this sense the intensity of a sound of given pitch is proportional to the velocity of the sound, to the density of the medium and to the square of the amplitude. In a formula the intensity is,

$$J = 2\pi^2 N^2 a^2 \delta V,$$

where $N =$ frequency, or number of complete vibrations per second,

$a =$ the amplitude of a vibration,
$\delta =$ density of medium and
$V =$ the velocity of propagation.

(Chwolson, *Traité de Physique,* Vol. I, Part 7, p. 905.)

INTENSITY OF SOUND. (3)

The intensity of sound is modified by the motion of the atmosphere and the direction of the wind and is strengthened by the neighborhood of a sonorous body.

(*Ganot's Physics, art.* 227.)

SOUND-INTENSITY IN TUBES.

In a speaking-tube the intensity of sound does not decrease with the square of the distance.

According to experiments carried on by Regnault, the distance to which a sound will carry in such a tube is roughly proportional to its diameter.

(*Ganot's Physics, art.* 229.)

EXPANSIONS AND CONTRACTIONS IN SOUND-TRANSMISSION ARE ADIABATIC.

In transmission of sound-waves, the expansions and compressions of the medium occur so rapidly that the expansions and compressions are adiabatic; namely, they occur in such manner that no heat is gained or lost to the volume of gas considered.

(Consult *Ganot's Physics, arts.* 231 and 507.)

VELOCITY OF SOUND, GENERAL PRINCIPLE.

The velocity of sound in any fluid equals the velocity acquired by a body in falling through one-half the height which represents the rate of variation of the pressure of the fluid with its density during a sudden change of density. Thus if V = velocity per second, g = acceleration of gravity, D = density of fluid and p = pressure, in the gravitational system of units,

$$V = \sqrt{g \frac{dp}{dD}}.$$

(Rankine, *The Steam Engine*, p. 321.)

VELOCITY OF SOUND IN AIR.

In dry air at 0° C. the velocity of sound is 331.7 meters, or 1088 feet, per second.

This velocity increases as the square root of the absolute temperature. Or, if T is the temperature in degrees centigrade, the velocity in meters per second at temperature T is,

$$V_T = 331.7\sqrt{1 + \frac{T}{273}}.$$

(Consult *Ganot's Physics, art.* 230. Also Chwolson, *Traité de Physique,* Vol. I, Part 7, p. 923, and *Smithsonian Physical Tables,* p. 102.)

VELOCITY OF SOUND AND AIR-DENSITY.

For the same temperature, the velocity of sound is independent of the density of the air and consequently of the pressure and is also roughly independent of the intensity and the quality of the sound.

(*Ganot's Physics, art.* 230.)

NEWTON'S FORMULA (modified by Laplace) FOR THE VELOCITY OF SOUND IN GASES.

The velocity of propagation of sound in a gas is directly as the square root of the elasticity of the gas and inversely as the square root of its density. The elasticity exceeds the isothermal elasticity P (measured by the pressure) by an amount γ which is the ratio of the specific heats of the gas at constant pressure and at constant volume. Calling ρ the density of the gas,

$$V = \sqrt{\frac{P\gamma}{\rho}}.$$

For air $\gamma = 1.41$ and the velocity of sound in air at 0° C. is 331.7 meters per second.

(Consult *Ganot's Physics, art.* 231. Also Chwolson, *Traité de Physique,* Vol. 1, Part 7, pp. 922–924.)

DOPPLER'S PRINCIPLE.

When a sounding body approaches the ear the note perceived is higher than the true one, but if the source recedes from the ear, the note perceived is lower.

If n = frequency of sounding body, V = velocity of compressional waves and v = velocity of body toward or from the ear, the pitch heard is $n' = \frac{nV}{V \mp v}$, where the minus sign is used for an approaching body and the plus sign for a receding body.

Doppler's principle may be extended to any system of waves in a medium.

(Ames, *Theory of Physics,* p. 160.)

VELOCITY OF SOUND IN LIQUIDS AND SOLIDS.

The rule of Newton that the velocity of propagation of sound equals the $\sqrt{\frac{\text{elasticity}}{\text{density}}}$ holds for liquids and solids as well as for gases.

For liquids the elasticity is the ratio of the pressure applied to the compression produced.

For solids Young's modulus may be taken as the value of the elasticity. The velocity, as in the case of gases, varies with the temperature.

(*Ganot's Physics, arts.* 234, 235. Also Chwolson, *Traité de Physique,* Vol. I, Part 7, pp. 929–933.)

REFLECTION OF SOUND-WAVES: ECHOES.

1. The angle of reflection is equal to the angle of incidence.
2. The incident sonorous ray and the reflected ray are in the same plane perpendicular to the reflecting surface.

An Echo is the repetition of a sound caused by its reflection from some surface transverse to its line of propagation.

(*Ganot's Physics, arts.* 236, 237.)

CHANGE OF PHASE AT REFLECTION.

When waves of sound pass from a less into a more dense medium, a portion of the energy is reflected back from the bounding surface without change of phase. If the waves are passing from a more into a less dense medium, the reflected wave undergoes a change in phase of one-half wave-length at the reflecting surface.

(Poynting and Thomson, *Sound,* pp. 104–108.)

PRINCIPLE OF RESONANCE.

One vibrating system may resonate, or be set into sympathetic vibration by another separate vibrating system when their natural periods of vibration are nearly equal. The more accurately they are tuned together the more marked is the resonance.

(Poynting and Thomson, *Sound,* pp. 58–62. Also Kimball, *College Physics,* pp. 204 *et seq.*)

REFRACTION OF SOUND. (1)

A sound-wave is refracted upon passing from a medium of one density into a medium of a different density. Sound may be deflected with a prism or focused with a lens. Sound travels poorly against the wind, because its wave-front is tilted upward.

(*Ganot's Physics, art.* 238.)

REFRACTION OF SOUND. (2)

When a sound-wave passes obliquely from one medium to another in which its velocity is different, its direction of propagation is changed. The laws of refraction are:

1. The normals to the incident and refracted wave-fronts and to the plane-surface all lie in the plane of incidence.
2. The ratio of the sine of the angle of incidence to the sine of the angle of refraction is constant for a given form of matter and waves of definite wave-number. It is entirely independent of the angle of incidence itself. $\frac{\sin i}{\sin r} = \mu$ is called the index of refraction.

(Consult Ames, *Theory of Physics,* pp. 424, 425. Also *Ganot's Physics, arts.* 238, 546, 547.)

INTERFERENCE OF SOUND.

Sound-interference can occur between two sonorous rays; the interference being determined by a difference in phase between the vibrations of the two wave-trains.

(Consult Chwolson, *Traité de Physique,* Vol. I, Part 7, pp. 949, 950.)

DIFFRACTION OF SOUND.

Diffraction phenomena are manifested more markedly with long than with short sound-waves and the length of sound-waves is such that there scarcely exists anything of the nature of a sound-shadow.

Lord Rayleigh has experimentally demonstrated the phenomena of the diffraction of sound.

(Chwolson, *Traité de Physique,* Vol. I, Part 7, pp. 954, 955.)

VELOCITY OF A TRANSVERSE WAVE ALONG A STRETCHED STRING.

The velocity of propagation of a transverse disturbance along a perfectly flexible stretched string or wire is given by

$$V = \sqrt{\frac{T}{m}}.$$

Here V may be taken as the velocity in cm. per second, the tension of the string in dynes and m its mass in grams per cm. of its length.

(Poynting and Thomson, *Sound,* pp. 93–95. Also, Kimball, *College Physics,* p. 216.)

TRANSVERSE VIBRATIONS OF A CORD.

1. The number of vibrations per second made by a cord under a given tension is inversely proportional to the length of the vibrating segment.
2. In case of two cords of equal length, and equal mass per unit length, the frequencies are proportional to the square roots of the tensions.
3. If two cords have equal lengths and are under equal tensions, their frequencies will be inversely proportional to the square roots of their masses per unit length.

Thus in a formula the frequency is,

$$N = \frac{1}{2l}\sqrt{\frac{T}{m}},$$

where l equals length of cord or the distance between two consecutive nodes, T equals the tension of the cord and m equals the mass per unit length.

(Kimball, *College Physics,* p. 216. Also *Ganot's Physics, art.* 268.)

NODES AND LOOPS IN AN ORGAN-PIPE.

In a closed organ-pipe the top is always a node or point of no vibration and the pipe, when the air column vibrates to the fundamental note, is one-quarter wave-length long. In the case of an open organ-pipe there is an antinode at each end and the pipe, when the air column vibrates to the fundamental note, is one-half wave-length long. Closing the end of a pipe lowers the tone one octave.

(*Ganot's Physics, art.* 275.)

NUMBER OF VIBRATIONS PRODUCED BY A MUSICAL PIPE.

When the length L of the pipe exceeds 12 times its diameter; for open pipes the frequency is,

$$n = \frac{p}{2L} V,$$

and for closed pipes

$$n = \frac{(2p-1)}{4L} V.$$

Here p is any whole number, as 1, 2, 3, etc., and V is the velocity of sound in air.

(*Ganot's Physics, art.* 277.)

LAW OF VIBRATION OF GEOMETRICALLY SIMILAR SYSTEMS.

When two vibrating systems are made of the same material and are geometrically similar but of different size their periods of vibration are in the same ratio as their linear dimensions.

(Kimball, *College Physics,* p. 230.)

VIBRATION OF RODS AND PLATES.

The number of transverse vibrations made in a given time by rods and thin plates of the same material is directly as their thickness and inversely as the square of their length.

(*Ganot's Physics, art.* 283.)

VIBRATION OF PLATES.

In plates of the same kind and shape, and giving the same system of nodal lines, the number of vibrations in a second is directly as the thickness of the plates and inversely as their area.

(*Ganot's Physics, art.* 284.)

VIBRATION OF BELLS.

Bells may be considered as curved plates. They do not vibrate as a whole, but for the fundamental they vibrate in four equal parts, these parts being separated by nodal lines. They are also capable of vibrating in 6, 8, 10 or 12 parts, producing thus a series of overtones. The note of a bell is higher in proportion as the surface is smaller and the substance thicker.

(*Ganot's Physics, art.* 284.)

ACOUSTIC ATTRACTION AND REPULSION.

The vibrations of an elastic medium attract bodies which are specifically heavier than itself and repel those which are specifically lighter. Thus in air a balloon filled with carbonic acid gas is attracted toward the opening in a resonance-box on which is a vibrating tuning fork, and a balloon filled with hydrogen is repelled.

(*Ganot's Physics, art.* 292.)

NUMERICAL VALUE OF PRESSURE OF SOUND.

Lord Rayleigh has shown that acoustic vibrations when they encounter the surface of a body must exert on it a pressure p, which for a plane-wave and normal incidence on a perfectly reflecting surface is,

$$p = \frac{2e}{V},$$

where e is the quantity of incident energy in unit time and V is the velocity of the sound.

(Chwolson, *Traité de Physique,* Vol. I, Part 7, p. 909.)

LIMITS OF AUDIBILITY.

The limits between which the frequencies of vibrations are audible vary considerably with different persons and the results of different investigators vary through rather wide limits. However, the lower limit seems to be between 16 and 24 vibrations per second, and the upper limit between 30,000 and 41,000 per second. Much depends also upon the intensity and quality of the sound as to whether or not it will be audible.

(*Ganot's Physics, art.* 244.)

AMPLITUDE NECESSARY TO MAKE SOUND-WAVES AUDIBLE.

It is found that sound-waves are inaudible if the amplitude of the sound-waves is less than about 8×10^{-8} cm. This limit is smaller as the pitch is higher.

(Poynting and Thomson, *Sound*, pp. 118, 119.)

COMBINATIONAL TONES.

When two pure tones are sounded simultaneously there is often heard, in addition to these two tones, two others. One of these has a pitch of frequency equal to the difference of the frequencies of the original tones, and the other has a pitch of frequency equal to their sum.

These tones are called Difference-Combinational tones and Summation-Combinational tones. They have been explained by Helmholtz.

(Helmholtz, *Sensations of Tone*. Also *Encyclopedia Britannica,* 10th Ed., Vol. XXV, p. 56.)

PURE TONE.

A pure tone is due to the disturbances sent out by a body vibrating with simple harmonic motion. Fourier has shown that any periodic disturbance may be made up of the resultant disturbances caused by a number of simple harmonic motions.

(Consult Helmholtz, *Sensations of Tone,* Chap. 1.)

LAW OF G. S. OHM.

"The human ear perceives pendular vibrations alone as simple tones, and resolves all other periodic motions of the air into a series of pendular vibrations, hearing the series of simple tones which correspond with these simple vibrations."

(Helmholtz, *Sensations of Tone*, p. 56.)

RESULTS OF VON HELMHOLTZ'S RESEARCHES.

Simple sounds, as those produced by a tuning fork with a resonance-box, are soft and agreeable, without roughness but weak and in deeper tones, dull. Musical tones accompanied by a series of harmonics, say up to the sixth, in moderate strength are full and rich. They are grander and more sonorous than simple tones. Such tones are produced by the pianoforte.

(These tone-qualities are very fully discussed in Chaps. IV and V of Helmholtz's *Sensations of Tone.*)

THE OCTAVE.

"A musical tone which is an octave higher than another, makes exactly twice as many vibrations in a given time as the latter."

(Helmholtz, *Sensations of Tone*, p. 13.)

EFFECT ON EAR OF A SYSTEM OF SOUND-WAVES.

"When several sonorous bodies in the surrounding atmosphere simultaneously excite different systems of waves of sound, the changes of density of the air, and the displacements and velocities of the particles of the air within the passages of the ear, are each equal to the (algebraical) sum of the corresponding changes of density, displacements, and velocities, which each system of waves would have separately produced, if it had acted independently."

(Helmholtz, *Sensations of Tone*, p. 28.)

ADDITION OF SIMPLE VIBRATIONS.

"Any given regular periodic form of vibration can always be produced by the addition of simple vibrations, having pitch-numbers which are once, twice, thrice, four times, etc., as great as the pitch-numbers of the given motion."

(Helmholtz, *Sensations of Tone*, p. 34.)

THE SUM OF PARTIAL TONES.

"Any vibrational motion of the air in the entrance to the ear, corresponding to a musical tone, may be always, and for each case only in one single way, exhibited as the sum of a number of simple vibrational motions, corresponding to the partials of this musical tone."

(Helmholtz, *Sensations of Tone*, p. 34.)

THE PRINCIPLE OF MUSICAL SCALES.

In the major, or diatonic scale, the frequencies of the notes bear the following ratios to that of the key note:

do re mi fa sol la si do

1, 9/8, 5/4, 4/3, 3/2, 5/3, 15/8, 2

This scale is built up of major triads, which have the pitch relation of do, mi, sol, or the frequency relation of 4, 5, 6. The minor scale is built up of minor triads, or notes with the frequency ratios 5, 6, 7½. The notes of this scale bear to the key note the relations

1, 9/8, 6/5, 4/3, 3/2, 5/3, 15/8, 2.

(Helmholtz, *Sensations of Tone*, p. 274.)

HELMHOLTZ'S THEORY OF CONSONANCE AND DISSONANCE.

When two notes are sounded simultaneously they produce an agreeable sensation in proportion as their frequencies form a simple ratio. Thus the octave, 1:2; the fifth, 2:3 and the fourth, 3:4 are the most consonant combinations, in the order named. A ratio 8:9 or 7:11 would be discordant. Helmholtz has shown that these facts are due to the absence of rapid beats in the case of consonant tones and their presence in the case of dissonant tones.

(Helmholtz, *Sensations of Tone,* pp. 228–330.)

IV

HEAT AND PHYSICAL CHEMISTRY

HEAT AND PHYSICAL CHEMISTRY

TEMPERATURE (Definition).

Temperature is a *condition of matter.* On the absolute thermo-dynamic scale, temperature is a quantity which is proportional to the mean kinetic energy E per molecule of the molecules of an "ideal gas," also to the product of the volume V and the pressure P of this gas. Thus,

$$T = KE = \frac{PV}{R},$$ where K and R are constants.

(Consult Maxwell, *Theory of Heat*, p. 51. Also Chwolson, *Traité de Physique,* Vol. III, Part 9, p. 7. Also article by E. F. Northrup, "High Temperature Investigation and a Study of Metallic Conduction." *Journal of the Franklin Institute,* June, 1915.)

QUANTITY OF HEAT (Definition).

Quantity of Heat is the total kinetic energy of the molecules, or ultimate particles of a body. Thus every store of heat-energy is expressed by the formula,

$$Q = \tfrac{1}{2} \Sigma mv^2 \cdot$$

In this formula the unit of heat energy is to be taken the same as the unit of mechanical energy, which is generally chosen on a system of absolute units and is the same as the unit of work. By m is to be understood the masses of the smallest particles of the body which are moving at any given instant with velocities v, different in general for each particle.

(Chwolson, *Traité de Physique,* Vol. III, Part 9, pp. 2 and 18.)

TEMPERATURE EQUILIBRIUM.

If two bodies A and B are in temperature equilibrium with a third body C, then A and B will be in temperature equilibrium with each other. It does not follow from the above that A, B, and C contain equal quantities of heat, even if they are all of equal mass and the same material. Thus A may be in grams of water at 0° C. and B in grams of ice at 0° C. Then A and B will be in temperature equilibrium with C, which is in grams of ice at 0 ° C., but A and B, though in temperature equilibrium with each other, contain different quantities of heat which differ by about 80 calories.

(Preston, *Theory of Heat,* Chap. I, Sec. II. See p. 20.)

EQUALIZATION OF TEMPERATURE.

When two bodies A and B are placed in contact, the temperature of the one body being higher than the temperature of the other body, the two bodies tend toward equality of temperature. The equalization occurs or tends to occur without oscillations of heat which have analogy with the oscillations of electricity, as observed when a condenser is discharged. The progress or the rate of equalization of temperature is a complex phenomenon which has relations with specific properties.

NEWTON'S LAW OF COOLING.

The rate at which a body cools is proportional to the excess of its temperature above the walls of the enclosure which surround it. Thus, $\log \frac{e}{E} = -a_1 t,$ which gives by differentiation, $-\frac{de}{dt} = ae.$ Here, e is the excess of temperature at time t, E the initial excess of temperature and a_1 and a are constants.

This law expresses the facts only approximately.

(Poynting and Thomson, *Heat,* p. 245. Also Preston, *Theory of Heat,* p. 528.)

DULONG AND PETIT'S CONCLUSIONS ON THE VELOCITY OF COOLING.

The cooling influence of gas surrounding a body is not affected by the nature of the surface of the body. The nature of the surface is effective only on the emissivity which would occur if the body were in a vacuum.

The empirical formula, which expresses the velocity of cooling V, is

$$V = k\,(a^{\theta} - a^{\theta_0}) + mp^{c}\,(\theta - \theta_0)\ 1.233.$$

(For discussion of principle stated and interpretation of formula, see Preston, *Theory of Heat*, pp. 530–540.)

ABSOLUTE (or LORD KELVIN'S) SCALE OF TEMPERATURE.

On the absolute, thermodynamic or Lord Kelvin's scale of temperature any two temperatures bear to each other the same ratio as the quantity of heat taken in at the higher temperature bears to the quantity of heat ejected at the lower temperature by a reversible engine working between the two temperatures as source and condenser. Thus,

$$\frac{T_1}{T_2} = \frac{Q_1}{Q_2}.$$

The efficiency of the perfect reversible engine is,

$$\frac{Q_1 - Q_2}{Q_1} = \frac{T_1 - T_2}{T_1}.$$

Here the heat quantities are expressed by Q and the absolute temperatures by T.

(Preston, *Theory of Heat*, p. 713. Also Rankine, *The Steam Engine*, p. 343.)

ABSOLUTE ZERO AND ABSOLUTE TEMPERATURE (Definitions).

The *absolute zero* (– 273.10° C.) on the gas-scale is the temperature at which an ideal gas would theoretically exert no pressure. It is numerically equal to the reciprocal of the pressure-coefficient a_v of the gas at constant volume.

The *absolute temperature* is the temperature reckoned from absolute zero. T (degrees absolute, now called degrees Kelvin) $= \frac{1}{a_v} + t$ (degrees centigrade).

(Consult *Smithsonian Physical Tables,* p. 247. Also *Ganot's Physics, arts.* 336, 337. Also paper by Arthur L. Day and Robert B. Sosman, "The Nitrogen Thermometer from Zinc to Palladium," *Amer. Jour. of Science,* Vol. XXIX, Feb., 1910. See pp. 100–102. Also Chwolson, *Traité de Physique,* Vol. III, Part 9, pp. 14–17.)

CARNOT'S THEOREM.

All reversible heat engines working between two given temperatures, and taking in and ejecting heat at the same two temperatures, have the same efficiencies. This efficiency is greater than that of any irreversible engine working between the same two temperatures.

(Poynting and Thomson, *Heat,* pp. 262–264; 265, 266.)

GAS-TEMPERATURE SCALE.

On the hydrogen or nitrogen gas-thermometer scale the temperature t, in degrees centigrade, is given by the relation

$$t = 100 \frac{P_t - P_o}{P_{100} - P_o},$$

where P_o is the pressure of the gas at 0° C., P_{100} its pressure at 100° C. and P_t its pressure at the measured temperature, the volume of the gas being maintained constant.

(Chwolson, *Traité de Physique,* Vol. III, Part 9, p. 23. See also p. 17.)

TEMPERATURE BY PLATINUM RESISTANCE-THERMOMETER.

Call $p_t = \frac{R_t - R_o}{R_{100} - R_o} 100$ the "platinum temperature," where R_t is the resistance of the thermometer at t degrees and R_0 and R_{100} its resistance at 0° and 100° on the centigrade scale. Then the difference between the true temperature and the platinum temperature is given by the formula of Callendar,

$$t - p_t = \delta\left[-\frac{t}{100} + \left(\frac{t}{100}\right)^2\right],$$

where δ is a constant, of value 1.50 for pure and greater for impure platinum.

(Northrup, *Methods of Measuring Electrical Resistance,* p. 298. Also, Burgess and LeChatelier, *Measurement of High Temperatures,* p. 197.)

EXPANSION OF BODIES WITH HEAT.

Nearly all bodies expand when they receive an additional quantity of heat. The expansion being slight, the coefficient of cubical expansion can with small error be taken equal to three times the coefficient of linear expansion.

For small ranges of temperature the expansion is very nearly proportional to the rise in temperature of the body.

(For experimental values, consult *Smithsonian Physical Tables,* pp. 232–235.)

EXPANSION COEFFICIENTS OF ANISOTROPIC BODIES.

1. The sum of the coefficients of linear expansion along any three directions mutually at right angles has a constant value equal to the sum of the three principal coefficients.
2. The coefficient of cubical expansion for an anisotropic body is equal to the sum of the coefficients of linear expansion along three directions mutually at right angles.

(Chwolson, *Traité de Physique,* Vol. III, Part 9, pp. 110, 111.)

EXPANSION OF LIQUIDS.

Liquids in general expand when they receive an additional quantity of heat, but water between 0° and 4° C. contracts, or increases in density, with increase in temperature or heat absorbed.

The real or absolute expansion of a liquid is the actual increase in volume, while the apparent expansion is that which is observed when a liquid contained in a vessel is heated, and this is less than the real expansion, because of the simultaneous expansion of the vessel itself.

(*Ganot's Physics, arts.* 322, 327, 331.)

DULONG AND PETIT'S LAW OF THERMAL CAPACITY.

For simple substances the atoms all have (approximately) the same thermal capacity, or the product of the specific heat by the atomic weight is the same for all elementary substances.

Regnault's mean value of this constant for 32 substances is 6.38.

(Preston, *Theory of Heat,* p. 294. Also *Ganot's Physics, art.* 464.)

NEUMANN'S LAW.

F. E. Neumann has found that the product of the molecular weight and specific heat remains constant for all compounds belonging to the same general formula and similarly constituted, but that the product varies from one series to another.

(Consult Preston, *Theory of Heat,* p. 296. Also *Ganot's Physics, art.* 465.)

HEAT-FLOW: LAW OF FLOW FOR STEADY STATE.

The quantity of heat Q which passes through a homogeneous solid enclosed between two parallel infinite planes at a distance d apart is expressed by,

$$Q = K\frac{\theta_1 - \theta_2}{d}At,$$

where θ_1 = the higher temperature of the one plane,

θ_2 = the lower temperature of the other plane,

A = the area through which the flow is reckoned,

t = the time the flow is measured and

K = a constant.

(Fourier, *The Analytical Theory of Heat*, Chap. I, Sec. IV. For the steady flow of heat between opposite faces of solids with certain geometric forms, see paper by Langmuir, Adams and Meikle in *Trans. of the Electrochem. Soc.*, Vol. XXIV, 1913, pp. 53–84. Also paper by E. F. Northrup, same Vol., pp. 85–106.)

GENERAL EQUATION FOR HEAT-FLOW.

The general differential equation for heat-flow is,

$$K\left(\frac{d^2\theta}{dx^2}+\frac{d^2\theta}{dy^2}+\frac{d^2\theta}{dz^2}\right)=c\frac{d\theta}{dt},$$

which, for the steady state, becomes

$$\frac{d^2\theta}{dx^2}+\frac{d^2\theta}{dy^2}+\frac{d^2\theta}{dz^2}=0.$$

Here, K = thermal conductivity,

c = thermal capacity per unit volume = specific heat × density.

θ = temperature and

t = time.

(Preston, *Theory of Heat, art.* 312. Also Fourier, *The Analytical Theory of Heat*, p. 112 *et seq.*)

STEADY FLOW OF HEAT FROM A POINT-SOURCE IN AN INFINITE ISOTROPIC MEDIUM.

1. The isothermal surfaces are concentric spherical shells with the point-source as center.
2. The flow of heat is perpendicular to the isothermal surfaces.
3. The total flow of heat across any isothermal surface is the same as that across any other, or any heat that is once within a tube of flow remains in it forever.
4. The flow per unit area through any cross-section of a tube of flow varies inversely as the area of the section, and hence inversely as the square of the distance from the point-source.

(Preston, *Theory of Heat, art.* 313.)

FLOW OF HEAT IN AN INFINITE CRYSTALLINE MEDIUM.

In an infinite crystalline medium, if heat be introduced at a single point, the isothermal surfaces, when the steady state is reached, will be a system of concentric and similar ellipsoids, the axes of any one of which are directly proportional to the square roots of the three principal conductivities of the crystalline medium.

(Preston, *Theory of Heat,* p. 675.)

FIRST LAW OF THERMODYNAMICS.

When work is done (namely, when measurable forces act through measurable distances or measurable electromotive forces give rise to measurable currents or measurable currents pass through measurable resistances, etc.), there is an equivalence between the work so done and the heat developed. This is expressed by the equation,

$$W = JH + w.$$

Here W is the total work done and H the heat developed. J is the equivalent of the work done in producing heat, and w is a quantity to express the processes occurring which cannot be measured as heat, such as the production of sound, radiant energy, etc. When w is zero,

W (in kilogram-meters) = 426.9 × H (in kilogram-calories).

Or (according to *Smithsonian Physical Tables*, p. 237)

W (in ergs) = 4.181 × 10^7 H (in, 20° C., gram-calories).

(Consult Hering, *Conversion Tables*, pp. 72 and 171. Also Preston, *Theory of Heat, art*. 37. For a precise statement, see Nernst, *Theoretical Chemistry*, pp. 7–10.)

SECOND LAW OF THERMODYNAMICS.

"Heat can never pass from a colder to a warmer body without some other change, connected therewith, occurring at the same time." (Clausius.)

(*Clausius on Heat*, p. 117. Also Preston, *Theory of Heat*, p. 49. For a scholarly treatment of the fundamental principles of thermodynamics, consult Chwolson, *Traité de Physique*, Vol. III, Part 9, Chap. VIII, pp. 409–550.)

SECOND LAW OF THERMODYNAMICS (Rankine's statement). (1)

"If the total actual heat of a homogeneous and uniformly hot substance be conceived to be divided into any number of equal parts, the effects of those parts in causing work to be performed are equal."

(Rankine, *The Steam Engine*, p. 306.)

SECOND LAW OF THERMODYNAMICS (Rankine's statement). (2)

"If the absolute temperature of any uniformly hot substance be divided into any number of equal parts, the effects of those parts in causing work to be performed are equal."

(Rankine, *The Steam Engine*, p. 307.)

TRANSFORMATION OF ENERGY.

"The effect of the presence in a substance, of a quantity of actual energy, in causing transformation of energy, is the sum of the effects of all its parts."

(This general law was first enunciated at the Phil. Soc. of Glasgow, Jan., 1853. Consult Rankine, *The Steam Engine,* p. 309.)

DIFFERENCE BETWEEN ABSORBED HEAT AND ENERGY.

The difference between the whole heat absorbed, and the whole expansive energy exerted in any thermic operation depends on the initial and final conditions of the substance alone in respect to pressure and volume and not on the intermediate process.

(Rankine, *The Steam Engine,* p. 304.)

CONVERTIBILITY OF ENERGY.

"All forms of energy are convertible. The total energy of any substance or system cannot be altered by the mutual actions of its parts."

(Rankine, *The Steam Engine,* p. 299.)

INTRINSIC ENERGY.

The total intrinsic energy of a body or system of bodies is never known. When bodies mutually react it is only the difference of the energy of each body in two states which is considered. If a body has less energy in its actual than in its standard state the expression for its energy is negative.

(Maxwell, *Theory of Heat,* p. 186. For a general treatment of the fundamental energy relations, consult Nernst, *Theoretical Chemistry,* pp. 7–10, 15–28.)

ENTROPY.

Entropy is a mathematical function introduced by Clausius. The entropy S of a substance is defined,

$$S = \int \frac{dQ}{T},$$

where dQ is an element of a quantity of heat and T is the absolute temperature of the element of the body which contains the element of heat dQ. In any change of condition of a body the change in its entropy is

$$S_1 - S_2 = \int_2^1 \frac{dQ}{T}$$

where S_1 is the value of the entropy in the final state of the body and S_2 the value in the original state of the body, and dQ is an element of heat gained by the body at absolute temperature T.

(For Clausius' original use of the term entropy see *Clausius on Heat,* p. 357. Rankine called the same quantity the "thermodynamic function." A clear explanation of the rather elusive meaning of the term entropy is to be found in Maxwell, *Theory of Heat,* pp. 162, 187, 189. Also Preston, *Theory of Heat,* pp. 724–726.)

CLAUSIUS' PRINCIPLE OF THE INCREASE OF ENTROPY.

All energy-changes in nature are such as to increase the total entropy of the universe. The entropy of the universe, therefore, tends toward a maximum. As the entropy increases the store of energy capable of transformation into useful work (available energy) diminishes, approaching a value zero. In this condition the entire energy of the universe will be in the form of heat and all bodies will be at the same temperature.

The above statement is a hypothetical extension to the *universe* of a principle recognized for the very limited portion which can be studied.

(Poynting and Thomson, *Heat,* p. 277 *et seq.*)

HEAT PRODUCED BY RADIUM.

Radioactive matter continually evolves heat. Curie and Laborde conclude from experiments that 1 gram of pure radium emits 100 gram-calories per hour. Thus 1 gram of radium emits per day nearly as much energy as will dissociate 1 gram of water. .

(Rutherford, *Radio-Activity*, p. 159.)

FUNDAMENTAL LAWS OF GASES.

Gases obey approximately the following laws:

1. *Boyle's* law, that for a constant temperature the volume of a gas diminishes in direct proportion to the pressure.
2. *Gay-Lussac's* law, that the volume of a gas at constant pressure increases proportionally with the absolute temperature.
3. *Avogadro's* law, that equal volumes of different gases at the same pressure and same temperature contain equal numbers of molecules.
4. *Dalton's* law, that the pressure of a mixture of several gases in a given space is equal to the sum of the pressures which each gas would exert by itself if confined in that space.
5. *Joule's* law, that gases in expanding do no interior work.

(For the laws which apply to a *perfect gas,* consult Chwolson, *Traité de Physique,* Vol. III, Part 9, Chap. IX, p. 551 *et seq.* Also Nernst, *Theoretical Chemistry,* pp. 38, 39.)

BOYLE'S (or MARIOTTE'S) LAW.

The temperature being constant, the volume of a given quantity of gas varies inversely as the pressure which it bears. Thus, $PV =$ a constant, where P is the pressure and V the volume of the gas.

This law, discovered by Boyle in 1662 and Mariotte in

France in 1679, is only approximately true for actual gases, and then only for low or medium pressures.

(*Ganot's Physics, art.* 181. Also, Chwolson, *Traité de Physique,* Vol. I, Part 3, p. 40 and Part 4, p. 422.)

VARIATIONS FROM BOYLE'S LAW.

1. No actual gas obeys Boyle's law rigorously. The divergence increases with the pressure.
2. Hydrogen is less compressible than Boyle's law requires; all other gases are more compressible.
3. The divergence from the law is greater for the easily liquefiable gases, such as carbonic acid, ammonia, etc., than for the gases formerly called permanent gases, oxygen, nitrogen, methane, etc.

(*Ganot's Physics, art.* 182. For a precise treatment of Boyle's and other gas-laws and their variations consult Nernst, *Theoretical Chemistry,* pp. 37–53.)

VAN DER WAALS' FORMULA.

Van der Waals has proposed as an accurate expression, relating the pressure and volume of a gas at any given temperature, the formula,

$$\left(p + \frac{a}{v^2}\right)(v - b) = \text{a constant.}$$

Here p is the pressure, v the volume and a and b are constants which differ for each gas.

(Consult Chwolson, *Traité de Physique,* Vol. I, Part 4, pp. 439–441. Also *Ganot's Physics, art.* 183.)

GAY-LUSSAC'S (or CHARLES') LAW.

The pressure of a gas being maintained constant its volume varies directly with the absolute temperature. If we define the coefficient of expansion a of a gas as the amount by which the unit of volume of the gas at 0° centigrade increases when the temperature is raised 1° C., the pressure being kept constant, then the law gives,

$$v_t = v_0 (1 + at),$$

where v_0 is the volume of the gas at 0° C. and v_t its volume at t° C. We may take for air a = 0.003665. The law of Charles holds for a wide variation in pressure.

(Maxwell, *Theory of Heat*, p. 29.)

VAN DER WAALS' FORMULA (combining the laws of Boyle and Charles, with corrections):

$$\left(p + \frac{a}{v^2}\right)(v - b) = RT.$$

Here, T = absolute temperature,
p = pressure,
v = volume and
R = gas-constant.
a and b = constants which differ for different gases.

When for the unit of pressure is taken a column of mercury of 1 meter and for the unit of volume, the volume of 1 kilogram of gas at 0° C. under a pressure of 1 meter of mercury, from Regnault's data:

for air a = 0.0037, b = 0.0026,
for CO_2 a = 0.0115, b = 0.003,
for H_2 a = 0.0000, b = 0.00069.

(Consult Chwolson, *Traité de Physique*, Vol. I, Part 4, Chap. II, see p. 441.)

ADIABATIC EXPANSION.

The law of expansion of a perfect gas, without receiving or emitting heat, gives for the relation between pressure p and volume v,

$$p = \frac{K}{v^{\gamma}}, \text{ where K is a constant}$$

For air, $\gamma = 1.4025$ and for steam in the perfectly gaseous state,

$$\gamma = 1.33.$$

(Rankine, *The Steam Engine*, pp. 319, 320. For values of γ, see *Smithsonian Physical Tables*, p. 243.)

ADIABATIC RELATIONS.

The adiabatic relations between the pressure p and the absolute temperature T, and between the volume v and the absolute temperature T, are given by the two relations,

$$Tp^{\frac{1-\gamma}{\gamma}} = \text{a constant, and}$$

$Tv^{\gamma-1}$ = a constant, where $\gamma = \frac{C_p}{C_v}$, the ratio of the specific heat at constant pressure to the specific heat at constant volume of the gas.

(Preston, *Theory of Heat*, p. 288.)

VARIATION OF PRESSURE WITH VOLUME IN A THERMALLY NON-CONDUCTING VESSEL.

The rate of variation of the pressure with the volume, when any fluid or gas is enclosed in a thermally non-conducting vessel, exceeds the rate of variation when the temperature is constant, in the ratio of the apparent specific heat of the fluid at constant pressure to its apparent specific heat at constant volume.

Symbolically expressed (according to Rankine):

$$\frac{dp}{dv} = -\gamma \frac{\frac{dp}{dT}}{\frac{dv}{dT}}, \text{ which becomes for a perfect gas,}$$

$$\frac{dp}{dv} = -\gamma \frac{p}{v}.$$

(Rankine, *The Steam Engine*, p. 320.)

EQUIPARTITION OF ENERGY: BOLTZMANN-MAXWELL LAW.

In a medium (a gas is usually considered) consisting of particles in motion the distribution of energy, throughout a given volume, will be such that, on the average, every mode of motion of its particles is equally favored, or, the kinetic energy is uniformly distributed among the degrees of freedom of the particles.

(Consult Campbell, *Modern Electrical Theory,* p. 229. For mathematical treatment, consult Jeans, *The Dynamical Theory of Gases,* pp. 67–69.)

LAW OF AVOGADRO.

When two gases are at the same pressure and temperature, the number of molecules in unit volume is the same for both.

Let N, m, $\bar{u}$ be the number of molecules, the mass of each molecule and the mean velocity of each molecule of the one gas, and N_1, m_1, $\bar{u}_1$ corresponding quantities for the other gas. By the kinetic theory of gases $\frac{1}{3}Nm\bar{u}^2 = \frac{1}{3}N_1m_1\bar{u}_1^2$ (where volume, pressure and temperature are the same), but $\frac{1}{2}m\bar{u}^2 = \frac{1}{2}\,m\bar{u}_1^2$ (by equipartition of energy). Hence the law that, $N = N_1$.

(Chwolson, *Traité de Physique,* Vol. I, Part 4, p. 491. Also Nernst, *Theoretical Chemistry,* pp. 200, 201.)

LAWS OF THE MIXTURE OF GASES (DALTON'S LAWS).

1. The mixture takes place rapidly and is homogeneous; that is, each portion of the mixture contains the two gases in the same proportion.
2. If the several gases and the mixture have the same temperature and if the several gases and the mixture occupy the same volume, then the pressure exerted by the mixture will equal the sum of the pressures exerted by the gases severally.

These laws are applicable to mixtures of gases and vapors.

(Preston, *Theory of Heat*, p. 71. Also *Ganot's Physics*, *art.* 388. Also Chwolson, *Traité de Physique*, Vol. I, Part 4, p. 468.)

JOULE'S LAW.

When an ideal gas expands in such a manner as *not to do any mechanical work* its temperature does not change.

Joule's experimental test of this law shows that no internal work is done by a gas during expansion, or in other words no molecular attractions have to be overcome.

Ordinary gases deviate slightly from Joule's law.

(Preston, *Theory of Heat*, p. 286. Also Nernst, *Theoretical Chemistry*, p. 42.)

SPECIFIC HEAT OF GASES.

It is concluded from experiments by Regnault and others that a gas has a specific heat which is independent of pressure in proportion as it approaches a perfect gas.

(Preston, *Theory of Heat*, p. 281. For discussion, and description of experiments, see Chwolson, *Traité de Physique*, Vol. III, Part 9, pp. 233–235.)

SPECIFIC HEAT OF A GIVEN VOLUME OF GAS.

The *difference* between the specific heats under constant pressure and under constant volume, referred to the unit of volume, is the same for all perfect gases taken at the same pressure and at the same temperature.

(Chwolson, *Traité de Physique*, Vol. III, Part 9, p. 220.)

LAW OF DELAROCHE AND BERARD.

This law states that for all elementary diatomic gases approximately in the perfect state, and for gaseous compounds formed without condensation and approximately in the perfect state, the product of the molecular weight and the specific heat at constant pressure has the same value.

(New Century Dictionary under word LAW. Also consult Nernst, *Theoretical Chemistry*, p. 42.)

X INTERNAL FRICTION OF A GAS.

It follows from the deductions of Maxwell that the internal friction or viscosity of any gas is a function of the absolute temperature but is independent of the density of the gas.

In respect to the density the statement is not rigorously exact for actual gases.

(Chwolson, *Traité de Physique,* Vol. I, Part 4, pp. 504–508. Also Poynting and Thomson, *Heat,* pp. 144–146.)

X THEOREM OF CORRESPONDING STATES.

If the pressure, volume and temperature of any gas at its critical point, be chosen for the unit values of these quantities, then (assuming its accuracy for representing the properties of a particular gas) Van der Waals' equation, with uniform values of the constants, will apply to all gases. In other words, all gases exhibit the same characteristics when at pressures and temperatures which are proportional to their critical pressures and temperatures.

(Edser, *Heat for Advanced Students,* pp. 312, 313. Also consult Nernst, *Theoretical Chemistry,* pp. 219–226.)

PRESSURE OF A GRAM-MOLECULE OF GAS.

The molecular weight of a chemical compound expressed in grams is called a *gram-molecule* or *mol.* (A gram-molecule or mol of O_2 is 32 grams, of H_2 2 grams and of H_2O 18 grams.)

The pressure exerted by one gram-molecule of any gas which closely obeys the gas-laws, when at 0° C. and when occupying a volume of one liter is 22.412 atmospheres.

(Nernst, *Theoretical Chemistry,* pp. 40, 41.)

THE GAS-CONSTANT.

For any gas which obeys the laws of an ideal gas

$$pv = \frac{p_0 v_0}{273} T = RT,$$

where p_0 and v_0 are the pressure and volume respectively of the gas at 0° C. and p and v its pressure and volume respectively at any absolute temperature T. The factor R, called the gas-constant, is conditioned only by the units of measurement chosen and is independent of the number of atoms in the molecule and the chemical composition of the gas.

If p is measured in atmospheres, v in liters and T in centigrade degrees reckoned from absolute zero, the value of R is 0.08204.

(Nernst, *Theoretical Chemistry,* p. 40. Also *Smithsonian Physical Tables,* p. 342.)

EQUATION OF CLAPEYRON.

For a perfect gas pv = RT, where p is the pressure and v the volume of the gas, T is the absolute temperature and R is a constant. The constant R is proportional to the mass M of the gas and for equal masses inversely proportional to the density δ. Thus,

$$R \propto \frac{M}{\delta},$$

(Chwolson, *Traité de Physique,* Vol. I, Part 4, pp. 437, 438.)

CONSTANT OF CLAPEYRON AND THE WORK OF EXPANSION OF A GAS.

The constant R in the equation, $pv = RT$, is numerically equal to the work of the expansion of the gas when its temperature is raised 1° C. under constant exterior pressure. Thus,

$$R = \frac{\gamma - 1}{\gamma} C_p J, \text{ where } \gamma = \frac{C_p}{C_v}$$

is the ratio of the specific heats at constant pressure and constant volume and J is the mechanical equivalent of heat.

(Chwolson, *Traité de Physique,* Vol. I, Part 4, pp. 493–495.)

FUNDAMENTAL EQUATION OF THE KINETIC THEORY OF GASES.

The fundamental equation of the kinetic theory of gases is,

$$pv = \tfrac{1}{3} Nm\overline{u}^2,$$

where v is the volume occupied by the gas, p the pressure of the gas, N the number of molecules contained in the volume v, m the mass of a molecule and $\bar{u}^2$ the mean square velocity of translation of the molecules.

(Chwolson, *Traité de Physique,* Vol. I, Part 4, p. 484. Also Preston, *Theory of Heat,* pp. 68–71. See serial article by Dr. Saul Dushmann in *General Electric Review,* Oct., Nov., Dec., 1915, on "The Kinetic Theory of Gases," Vol. XVIII, pp. 952–958, 1042–1049, 1159–1168.)

PRESSURE AND ENERGY OF GAS.

The kinetic theory of gases states that the pressure of a gas is equal to two-thirds the energy of translational motion of the molecules which are contained in the unit of volume of the gas, also that the energy of translational motion of the molecules of a gas is proportional to the absolute temperature of the gas. Thus, for the unit of volume,

$$p = \frac{2}{3} E_1 \text{ and } RT = \frac{2}{3} E_1,$$

where p is the pressure,

T the absolute temperature,

R a constant and

E_1 the energy of translational motion in the unit of volume.

Important relations of the kinetic theory of gases are:

$$pv = \frac{1}{3} Nm\bar{u}^2 = \frac{1}{3} M\bar{u}^2 = RT = \frac{2}{3} E.$$

(Chwolson, *Traité de Physique,* Vol. I, Part 4, pp. 489, 490.)

VELOCITIES OF GAS-MOLECULES.

1. The velocity of the molecules of a given gas is proportional to the square root of the absolute temperature of the gas.
2. The velocities of the molecules of different gases, at the same temperature, are inversely proportional to the square roots of the densities of these gases. Or in a formula,

$$u = \sqrt{3gR_0 \frac{T}{\delta}},$$

where g and R_0 are constants, T absolute temperature and δ the density of the gas.

(Chwolson, *Traité de Physique,* Vol. I, Part 4, p. 490.)

MAXWELL'S LAW OF MOLECULAR VELOCITIES.

The components of molecular velocity (in a gas) are distributed among the molecules according to the same law as the errors are distributed among the observations in the theory of errors of observations.

(Preston, *Theory of Heat,* pp. 71, 72.)

WORK PERFORMED WHEN TWO GASES MIX.

When two gases which exhibit no chemical interaction become mixed by diffusion of each into the other no work is performed if the volume of mixed gases remains constant. If, on the other hand (with an arrangement which may be realized experimentally), the volume of the first gas increases from V_1 to $V_1 + V_2$ and the volume of the second gas increases from V_2 to $V_2 + V_1$, then the first gas in diffusing will do external work:

$W_1 = n_1 RT \log_e \frac{V_1 + V_2}{V_1}$, and the second gas will do external work $W_2 = n_2 RT \log_e \frac{V_1 + V_2}{V_2}$.

Here T is the absolute temperature, R the gas-constant, and n_1 and n_2 are the number of gram-molecules concerned in the diffusion of the first and second gases respectively.

The total external work done by the mixing of the two gases is,

$$W = W_1 + W_2 = RT\left(n_1 \log_e \frac{V_1 + V_2}{V_1} + n_2 \log_e \frac{V_1 + V_2}{V_2}\right).$$

This formula, developed by Rayleigh and more thoroughly by Boltzmann, expresses a law which holds good universally.

(Nernst, *Theoretical Chemistry,* pp. 96–100.)

NUMBER OF MOLECULES IN A GAS.

It is deduced from theory that, in a cubic centimeter of air, or (according to the law of Avogadro) in every other gas, there are contained in a gram-molecule of the gas about 4.5×10^{23} molecules.

(Chwolson, *Traité de Physique,* Vol. I, Part 4, p. 511.)

LAWS OF ABSORPTION OF GASES BY LIQUIDS: HENRY'S LAW.

1. For the same gas, the same liquid and the same temperature, the *weight* of gas absorbed is proportional to the pressure, or, at all pressures, the *volume* dissolved is the same. (Known as Henry's law.) The volume absorbed varies with the gas. Thus, water dissolves over fifty thousand times as great a volume of ammonia as of nitrogen. The absorbing power also varies with the liquid. Thus alcohol absorbs gases better than water.
2. The quantity of gas absorbed decreases with increase of temperature.
3. The quantity of gas which a liquid can dissolve is independent of the nature and of the quantity of other gases which it may already hold in solution.

1 and 3 are only rigorously exact for gases which are but slightly soluble and when the pressures do not exceed a few atmospheres.

(*Ganot's Physics, art.* 190. Also Walker, *Introduction to Physical Chemistry,* pp. 57, 58.)

SOLUTION IN LIQUID OF MIXED GASES: DALTON'S LAW.

When a mixture of gases dissolves in a liquid, each component of the mixture dissolves proportionally to its own partial pressure; or when a liquid acts to dissolve mixed gases, each gas dissolves as if all the others were absent. (Known as Dalton's law.)

Both Dalton's and Henry's law hold well only when the gases are slightly soluble and the pressures do not exceed a few atmospheres. The divergencies are large for very soluble gases and great pressures.

(Walker, *Introduction to Physical Chemistry,* p. 58.)

LAW OF PARTIAL PRESSURE.

In a mixture of liquids the vapor emitted by the liquids will in general have the same components as the liquid mixture remaining behind. The ingredients exert *partial pressures* the sum of which is the vapor-pressure of the mixture. The law holds good universally and is of fundamental importance, that: the partial pressure of each component of a mixture of liquids is always less than the vapor-pressure of a component in the free or unmixed state, the temperature being the same.

(Nernst, *Theoretical Chemistry,* pp. 105–107.)

WORK DONE BY EVAPORATION.

A liquid which evaporates against the constant external pressure of its saturated vapor performs external work and absorbs heat. The external work performed by the evaporation of one gram-molecule of any simple liquid is independent of the nature of the liquid and is directly proportional to the absolute temperature T at which the evaporation takes place. Thus the external work performed in the evaporation of one gram-molecule of *any* liquid is,

$$W = p(V - V') = RT - pV'.$$

Here p is the external pressure and V the volume of the vapor. V′ is the volume of the one gram-molecule of the liquid before evaporation started and R is the gas-constant. When, as usual V′ is negligible and p is in atmospheres and V in liters we have,

$$W = 0.0821T \text{ liter-atmospheres.}$$

(Nernst, *Theoretical Chemistry,* p. 57.)

ABSORPTION OF GASES BY SOLIDS.

The surfaces of solids by exerting an attraction on the molecules of gases become covered with a layer of *condensed* gas, and porous solids, which present an extended surface to the gas, tend to absorb it, many solids absorbing a considerable quantity of gas. This absorption takes place without any chemical change. The absorption is in general greater in the case of the more easily liquefiable gases.

(*Ganot's Physics, art.* 194.)

OCCLUSION OF GASES.

At a high temperature, platinum and iron allow hydrogen to traverse them quite readily. Some metals will absorb gases when cooling and give them off when heating. This property is most marked with palladium, which not only absorbs hydrogen while being cooled after being heated but even when cold. This may cause a palladium wire to lengthen quite perceptibly.

(*Ganot's Physics, art.* 195.)

LAW OF DIFFUSION OF GASES (GRAHAM'S LAW).

The quantity of a gas which passes through a porous diaphragm in a given time is inversely as the square root of the molecular weight of the gas (or its density).

(*Ganot's Physics, art.* 191. Also consult Poynting and Thomson, *Heat,* p. 327.)

EFFUSION OF GASES.

When gas passes through a small aperture, about 0.013 mm. in diameter, from a region where its pressure is h (expressed in terms of the height of a column of the gas which would exert the same pressure as that of the effluent gas), into a region where its pressure is h′ the velocity of efflux, or the rate of *effusion,* is,

$$v = \sqrt{2g\,(h-h')},$$

where $g =$ the acceleration of gravity.

Or we can say: the velocities of efflux, or the rates of effusion of various gases, are inversely as the square roots of their densities.

(*Ganot's Physics, art.* 192.)

BOILING.

1. The temperature of ebullition, or the boiling-point, increases with the pressure.
2. For a given pressure boiling begins at a certain temperature, which varies for different liquids, but which, for equal pressures, is always the same in the same liquid.
3. Whatever be the rate of input of heat into the liquid, as soon as boiling begins the temperature of the liquid remains stationary.

(*Ganot's Physics, art.* 366. For cases of "Superheating" see Preston, *Theory of Heat,* p. 360. Also consult Nernst, *Theoretical Chemistry,* pp. 63, 64.)

BOILING AND VOLATILIZATION.

If a substance is liquid at a temperature at which the pressure of its vapor equals the pressure on its surface, the substance will liquefy and boil, but, if the substance is solid at a temperature at which its vapor has the pressure of the pressure on its surface, the substance changes directly from a solid to a vapor; namely, it will volatilize or sublime.

Water is an example of the first, arsenic and carbon are examples of the second.

(See Preston, *Theory of Heat,* p. 370. Also Walker, *Introduction to Physical Chemistry,* p. 82. Also Nernst, *Theoretical Chemistry,* pp. 70, 476.)

LATENT HEAT OF VAPORIZATION: TROUTON'S LAW.

For different liquids the latent heat of vaporization multiplied by the molecular weight is approximately proportional to the absolute temperature at which vaporization occurs; or the molecular latent heat is approximately proportional to the absolute temperature.

Thus calling w the molecular weight of the vapor, L the latent heat of vaporization of the liquid and T the absolute temperature,

$$\frac{wL}{T} = \text{a constant.}$$

(Preston, *Theory of Heat*, p. 391. Also consult Nernst, *Theoretical Chemistry*, pp. 272–274.)

VAPOR-PRESSURE IN COMMUNICATING VESSELS AT DIFFERENT TEMPERATURES.

"When two vessels containing the same liquid, but at different temperatures, are connected, the pressure is identical in both vessels, and is the same as that corresponding to the lower temperature."

The liquid distils from the vessel at higher temperature to the vessel at lower temperature.

(*Ganot's Physics, art.* 364.)

VAPOR-PRESSURE OF MIXED LIQUIDS.

1. Liquids which do not mix: the vapor-pressure equals the sum of the vapor-pressures of the constituents.
2. Liquids which mix partially: the vapor-pressure is less than that of the sum of the pressures of the constituents.
3. Liquids which mix in all proportions: the diminution of vapor-pressure is still greater.

(Preston, *Theory of Heat*, p. 406.)

VAPOR FORMATION IN A VACUUM.

If any simple volatile liquid is brought into a vacuous space the liquid evaporates with great rapidity and evaporation continues until the vapor formed exerts a certain definite maximum pressure. If, when this pressure is reached, some liquid still remains in the space originally vacuous, the maximum pressure obtained will be the so-called *vapor-pressure* of the liquid at the particular temperature of the experiment. When the temperature increases the vapor-pressure increases, and usually very rapidly.

(Nernst, *Theoretical Chemistry*, pp. 56, 57. Also *Ganot's Physics, art.* 355.)

CONDENSATION OF SATURATED VAPOR.

If air be perfectly free from dust particles it may be considerably supersaturated without condensation in the form of a cloud.

If air be ionized condensation is accelerated, the negative ions being more effective nuclei than the positive ions. This fact has been utilized in counting the number of ions in a gas.

(Poynting and Thomson, *Heat,* pp. 168–172. Also Thomson, *Conduction of Electricity Through Gases,* pp. 163–187.)

VAPOR-PRESSURE.

The pressure of a vapor in contact with its own liquid depends only upon the temperature and is independent of the relative proportions of the liquid and vapor.

The pressure is some function of the temperature but *not* a linear function.

When the pressure upon the surface of a liquid is due to the atmosphere, the liquid will boil at the moment when its vapor-pressure just exceeds that of the atmosphere. A diminution of 1° C., in the boiling-point of water, corresponds to an ascent of about 1080 feet.

(Preston, *Theory of Heat,* pp. 148 and 408.)

LAW OF MASS-ACTION (Credited to Guldberg and Waage).

As stated by the authors of the law, the rate of chemical action is proportional to the *active mass* of each of the reacting substances. It may be thus stated: when any substance in solution enters into a chemical reaction, the amount of reaction in the unit of time is proportional to the active mass of the substance, namely, to the number of gram-molecules of the substance contained in unit volume of the solution.

(See Walker, *Introduction to Physical Chemistry*, p. 277 *et seq.* Also consult Nernst, *Theoretical Chemistry*, pp. 443–446, for an analytical treatment of this very fundamental law of *chemical kinetics.*)

LAW OF RELATIVE PROPORTIONS IN EQUILIBRIUM.

The condition of physical or chemical equilibrium in a heterogeneous system is independent of the relative mass of each phase present in the system.

Thus, at a given pressure and temperature the *physical* equilibrium between water and its vapor is undisturbed by an increase in the mass of either phase; and the *chemical* equilibrium between $CaCO_3$,CaO and CO_2 is undisturbed by a change in the quantity by weight of any of the substances enumerated.

(Nernst, *Theoretical Chemistry*, p. 471 *et seq.*)

LAW OF DISTRIBUTION, AMONG SEVERAL MOLECULAR SPECIES.

When several molecular species [as acetic acid, giving in both vapor form and in solution the single molecules CH_3CO_3H and the double molecules $(CH_3CO_3H)_2$] evaporate at constant temperature from a common solvent (as benzene) into a fixed vapor-space the ratio of the concentration in the *vapor-space*, of any one molecular species, to its concentration in the *solvent* is constant.

This distribution in the quantity of a molecular species between the solvent and the vapor-space is independent of the presence of other molecular species, even when these latter are chemically reactive with the former.

(Nernst, *Theoretical Chemistry*, p. 491. For application of the "distribution law" to dilute solutions see Washburn, *Principles of Physical Chemistry*, pp. 148–150.)

THE LAW OF HESS.

The total quantity of heat, disengaged in the passage of a group A of substances to a group B, is independent of the nature of this passage, namely, of the character and of the order of intermediate reactions, provided the physical state (in the wide sense of this word) of the groups A and B is the same in all cases.

This law is a basic principle of thermochemistry.

(Consult Chwolson, *Traité de Physique,* Vol. III, Part 9, p. 284.)

EFFECT OF TEMPERATURE ON BALANCED CHEMICAL ACTION.

If a direct chemical action gives out a certain quantity of heat per gram-molecule transformed, the reverse reaction will absorb an exactly equal quantity of heat; and rise of temperature always affects chemical equilibrium in such a manner that the displacement of the point of equilibrium takes place in the direction which will determine absorption of heat.

(See Walker, *Introduction to Physical Chemistry,* p. 293. Also Nernst, *Theoretical Chemistry,* pp. 673–676. Note statement of LeChatelier on p. 676.)

PROGRESS OF CHEMICAL DECOMPOSITION.

The phenomena of chemical decomposition of a body in a confined space go on, if one of the elements of the decomposition is gaseous, until a certain pressure is attained when, for a particular temperature, the decomposition ceases.

Deville used the word "Dissociation" which has analogy under the above conditions with vaporization of liquids.

(Roscoe and Schorlemmer, *Treatise on Chemistry,* Vol. II, pp. 129–132.)

CRITICAL TEMPERATURE.

There is for every substance a *critical temperature* above which the substance cannot be liquefied with pressure; or there is for each gas a particular or critical temperature to which the gas must be cooled in order to liquefy it with any pressure.

(Preston, *Theory of Heat*, p. 450 *et seq.* See CO_2 curves, pp. 490, 491. Also consult Nernst, *Theoretical Chemistry*, pp. 64–66.)

✗ MOLECULAR SURFACE-ENERGY: LAW OF EÖTVÖS.

According to Eötvös, the work required to form the surface of a spherical gram-molecule of a liquid varies with the temperature in the same manner for all liquids. This law is stated in the expression,

$$\gamma v^{2/3} = k(T - T_0),$$

where v is the volume occupied by one gram-molecule of a liquid and γ is its surface tension. T_0 is a temperature taken not far from the critical, T the temperature of the liquid, and k is a constant, independent of the nature of the liquid. The quantity $\gamma v^{2/3}$ is proportional to the molecular surface-energy of a sphere formed from one gram-molecule of the liquid in question.

The law of Eötvös has importance in determining the molecular weight of liquids.

(Nernst, *Theoretical Chemistry*, pp. 275–277.)

CURIE'S LAW.

In paramagnetic substances the magnetic susceptibility, or the ratio of the intensity of magnetization to the magnetizing force, is inversely proportional to the absolute temperature.

(Richardson, *The Electron Theory of Matter*, p. 378.)

LAWS OF DIFFUSION IN LIQUIDS.

1. The rate at which the diffusion of any substance goes on is proportional to the rate of variation of the strength of that substance in the fluid as measured along the line in which the diffusion takes place.
2. The rate of diffusion varies with the temperature.

The law of diffusion of matter has exactly the same form as that of the diffusion of heat by conduction.

(See Maxwell, *Theory of Heat,* Chap. XIX, and p. 276. Also consult *Ganot's Physics, art.* 140. Also Nernst, *Theoretical Chemistry,* p. 151 *et seq.*)

OSMOSE, OSMOSIS OR DIOSMOSE.

When two liquids which will mix are separated only by a porous membrane there is a movement of the liquids in both directions through the membrane. The greater movement is usually from the less dense to the more dense liquid so as to cause the level of the more dense liquid to rise above that of the less dense. This action increases with the temperature and is proportional to the strength of the solution.

(*Ganot's Physics, art.* 139. Also Chwolson, *Traité de Physique,* Vol. I, Part 5, pp. 661–668. Also Nernst, *Theoretical Chemistry,* pp. 125–127.)

OSMOTIC PRESSURE AN ANALOGUE OF GAS-PRESSURE.

A principal feature in the analogy between a dissolved substance and a gas consists in the correspondence between the energy content of each, this energy content being independent, at any fixed temperature, of the volume occupied by a given mass of either. Thus, the osmotic pressure of a dissolved substance is exactly the same as the gas-pressure which would be exerted if the solvent were removed and the dissolved substance in gaseous form were left behind to occupy the same volume at the same temperature. Thus the gas-law for a dissolved substance may be written $PV = RT = 0.0821T$ liter-atmospheres.

Here P denotes the osmotic pressure in atmospheres of a solution which contains one gram-molecule of the substance dissolved in V liters of solvent, and T the absolute temperature in degrees centigrade.

(Nernst, *Theoretical Chemistry*, p. 144.)

RAOULT'S LAW ON THE LOWERING OF VAPOR-PRESSURE.

"The relative lowering of vapor-pressure experienced by a solvent on dissolving a foreign substance is equal to the quotient obtained by dividing the number of dissolved molecules n, by the number of molecules N, of the solvent." Thus in symbols

$$\frac{p - p'}{p'} = \frac{n}{N}.$$

(Nernst, *Theoretical Chemistry*, pp. 144, 145. See also p. 263 *et seq.*)

GAS-LAWS APPLIED TO SOLUTIONS.

1. The osmotic pressure is, at constant temperature, proportional to the concentration of the solution, or inversely proportional to the volume occupied by a given quantity of the dissolved substance (analogue of Boyle's law).
2. The osmotic pressure is proportional to the absolute temperature (analogue of Gay-Lussac's or Charles' law).
3. Equal volumes of isotonic solutions—solutions which exercise the same osmotic pressure—when under the some pressure and at the same temperature contain the same number of molecules. This number of molecules is the same as that of a gas under like conditions of volume, pressure and temperature (analogue of Avogadro's law).

(Chwolson, *Traité de Physique*, Vol. I, Part 5, p. 666. See also Walker, *Introduction to Physical Chemistry*, p. 184. Also Nernst, *Theoretical Chemistry*, p. 141 *et seq.*)

LAW OF KOHLRAUSCH FOR DILUTE SALT-SOLUTIONS.

The *molecular conductivity* of a solution (namely, its electrical conductivity times the volume of the solution which contains one gram-molecule of the dissolved substance) is independent of the concentration of the solution for very dilute salt-solutions. This constant value of the molecular conductivity of an electrolyte, at infinite dilution, is the sum of two numbers, one of which depends solely upon the speed of migration of the positive ions and one solely upon the speed of migration of the negative ions.

(Walker, *Introduction to Physical Chemistry*, pp. 251, 256. Also Nernst, *Theoretical Chemistry*, pp. 365–367.)

ADDITIVE PROPERTY OF DILUTE SOLUTIONS.

In a sufficiently dilute electrolytic solution there is a complete dissociation into ions of the dissolved substance and it is a fundamental law that: "the properties of a salt-solution are composed additively of the properties of the free ions."

The law, to hold true, presupposes *complete* dissociation.

(Nernst, *Theoretical Chemistry*, p. 384.)

SIMILARITIES IN BEHAVIOR OF IONS AND MOLECULES.

It may be stated, as a theorem, that: the *ions* of dissociated substances in solution exhibit all the properties of neutral molecules and some additional properties due solely to the electrical charge of the former. Thus ions, like molecules, diffuse, exert pressure, distribute according to the law of equipartition of energy, show additive properties, etc., and in addition, in virtue of their electric charge, transport electricity.

(Nernst, *Theoretical Chemistry*, pp. 392, 393.)

OSTWALD'S LAW OF MOLECULAR CONDUCTIVITY.

This law is expressed by the formula,

$$\frac{m^2}{(1-m)v} = k, \text{ a constant.}$$

k is called the dissociation constant, v is the dilution and m represents the degree of ionization and is equal to

$$\frac{\mu_\nu}{\mu_\infty},$$

where μ_ν is the molecular conductivity at dilution ν and μ_∞ is the molecular conductivity at infinite dilution.

(Walker, *Introduction to Physical Chemistry*, Chap. XXV, see pp. 262–265.)

FORMATION OF "HYDRION."

Aqueous solutions of various acids possess one property in common, on the dissociation theory: they form "hydrion," namely, yield hydrogen atoms in the ionic state.

The electrical conductivity of weak solutions of acids is due almost wholly to the hydrion they contain, and measuring the relative strengths of acids by measuring the conductivity of the solutions has practically superseded all other methods, especially for the weaker organic acids.

(Walker, *Introduction to Physical Chemistry*, p. 313.)

PROCESS OF NEUTRALIZATION.

It is shown by conductivity measurements of nearly pure water that hydrogen ions and hydroxyl ions can only exist beside each other in mere traces, if at all. Hence when we bring together in a water-solution two electrolytes (an acid and a base) which yield hydrogen ions and hydroxyl ions these unite to form water according to the equation

$$\overset{+}{H} + O\overline{H} = H_2O.$$

This process is called *neutralization*.

(Nernst, *Theoretical Chemistry*, pp. 517, 518.)

NEUTRALIZATION OF AN ACID AND A BASE.

When a strong acid is mixed with a strong base the *hydrogen ions* and the *hydroxyl ions* unite almost completely to form *molecules of water* and the general result follows, that: "the neutralization of a strong acid by a strong base must always exhibit the same heat of reaction"—about 13,700 gram-calories.

(See Nernst, *Theoretical Chemistry*, pp. 578–610. Also Washburn, *Principles of Physical Chemistry*, p. 243. Also see for values of heats of neutralization, *Smithsonian Physical Tables*, p. 213.)

RELATIVE AVIDITY OF ACIDS.

Two weak acids share between themselves a base for which both are competing in a certain definite ratio. The ratio of the avidities of the two acids is (under specified conditions) equal to the square root of the ratio of their dissociation-constants. Thus,

$$\frac{x}{1-x} = \sqrt{\frac{k}{k'}},$$ where x is the amount of one of the two acids neutralized by the base and k and k′ are the dissociation-constants of the two acids.

(Walker, *Introduction to Physical Chemistry*, pp. 314–316. Also Nernst, *Theoretical Chemistry*, pp. 521–523.)

CHEMICAL COMBINATION OF THE ELEMENTS.

"Combination always takes place between certain definite and constant proportions of the elements or between multiples of these."

The composition of a chemical compound can be learned by (1) bringing the component elements into combination under favorable conditions: *synthesis*, or (2) by separating it into its component elements: *analysis*.

(Roscoe and Schorlemmer, *Treatise on Chemistry*, Vol. I, p. 72. Also Nernst, *Theoretical Chemistry*, pp. 30, 31.)

PERIODIC SYSTEM OF CHEMICAL ELEMENTS.

The properties of the chemical elements are periodic functions of their atomic weights.

This is shown by means of a systematic arrangement of the elements in a table known as "Mendelejeff's Periodic System of the elements."

(See table, Nernst, *Theoretical Chemistry*, p. 180. Also see table published July, 1915, in *General Electric Review*, by the Research Laboratory of the General Electric Company.)

LAW OF GAY-LUSSAC AND HUMBOLDT ON COMBINATIONS OF GASES BY VOLUME.

The volumes in which gaseous substances combine chemically bear a simple relation to one another and to the volume of the resulting product.

(Roscoe and Schorlemmer, *Treatise on Chemistry*, Vol. I, p. 75.)

RICHTER'S LAW.

When any two neutral salts undergo double decomposition by interchange of their acid and basic constituents the two new salts resulting from such interaction are also neutral in character.

(Roscoe and Schorlemmer, *Treatise on Chemistry*, Vol. I, p. 33.)

"MOLECULAR ROTATION" OF AN OPTICALLY ACTIVE LIQUID.

A plane-polarized ray in passing through a liquid optically active is rotated proportionally both to the length of the path of the ray through the liquid and to the density of the liquid. Hence when the actual observed rotation is divided by the density and the length of the path the *specific rotary power* of the liquid is obtained, and when this is multiplied by the molecular weight the *molecular rotation* is expressed.

(Walker, *Introduction to Physical Chemistry*, p. 158.)

OUDEMAN'S LAW OF OPTICAL ROTATION.

The molecular rotation of plane-polarized light by salts of an optically active acid or base always tends to a definite limiting value as the concentration of the solution diminishes. This regularity is known as Oudeman's law.

(Walker, *Introduction to Physical Chemistry,* p. 174. Also Nernst, *Theoretical Chemistry,* p. 391.)

RAOULT'S LAW FOR DEPRESSION OF FREEZING-POINT.

If the same number of molecules of different substances be dissolved in a given number of molecules of the solvent the depressions of the freezing-points of the solutions are equal.

(New Century Dictionary under word LAW. Also Walker, *Introduction to Physical Chemistry,* p. 210 *et seq.* Also Nernst, *Theoretical Chemistry,* p. 146.)

INFLUENCE OF PRESSURE ON THE MELTING-POINT.

The melting-point of a substance (as ice) which contracts on liquefaction, is lowered by increase of pressure and if the substance expands during liquefaction the melting-point is raised by increase in pressure.

(Preston, *Theory of Heat,* pp. 341, 342. Also Nernst, *Theoretical Chemistry,* p. 67 *et seq.*)

DISSOLVED SALTS RAISE BOILING-POINT.

The boiling-point of a liquid is raised by dissolving in it a salt. Thus, a saturated solution of common salt in water boils at about 102° C. and water saturated with calcium chloride boils at about 179° C.

The boiling-point is lowered when a gas is dissolved in the liquid.

(Walker, *Introduction to Physical Chemistry,* pp. 195, 196. Also *Ganot's Physics, art.* 368.)

FREEZING-POINT OF A SOLUTION (BLAGDEN'S LAW).

The freezing-point of a substance is lowered by dissolving in it some foreign matter. The change is usually proportional to the amount of dissolved-substance in the solvent, although sometimes the change is abnormally great. This is the case with substances which have an abnormal osmotic pressure.

(Consult Walker, *Introduction to Physical Chemistry*, pp. 65, 195 and 406.)

FUSION: OF CRYSTALLINE SOLIDS AND PURE METALS.

1. For a given pressure the temperature of fusion is fixed, and is the same as that of solidification.
2. While fusion or solidification is taking place the temperature of the whole mass remains constant.
3. During fusion heat is absorbed by the substance and an equal quantity of heat is disengaged during solidification.

(Preston, *Theory of Heat*, p. 336.)

THE CRYOHYDRIC TEMPERATURE.

The *cryohydric* temperature is the lowest temperature that can be produced by mixing salt and ice together. No solution of salt in water can exist in a stable state below this temperature.

(For full and accurate information on this matter, consult Walker, *Introduction to Physical Chemistry*, Chap. VIII.)

GIBBS' CRITERIA OF THERMAL EQUILIBRIUM.

For the equilibrium of any isolated system it is necessary and sufficient that, if the total energy of the system be constant, the variation of its entropy shall either vanish or be negative; or, if the total entropy of the system be constant, the variation of its energy shall either vanish or be positive.

(Gibbs, "On the Equilibrium of Heterogeneous Substances," *Trans. Connecticut Acad.*, Vol. 3, p. 109.)

THE PHASE-RULE.

The Phase-Rule, due to Gibbs, generalizes the conditions which determine the equilibrium of any system. Let K represent the number of components (salt, water, etc.) and let i be the total number of phases in which these components are present (crystalline, liquid, vapor, etc.). The number of degrees of freedom, or independent ways in which the system can be changed, is given by

$$n = K + 2 - i.$$

(*New Century Dictionary* under words PHASE RULE. Also Jones, *Elements of Physical Chemistry*, p. 489 *et seq.* Also Walker, *Introduction to Physical Chemistry*, Chap. XI, pp. 103–117. Also Gulliver, *Metallic Alloys*, pp. 157–164.)

LAW OF THE MUTUALITY OF PHASES.

"If two phases, respecting a certain definite reaction, at a certain temperature, are in equilibrium with a third phase, then at the same temperature and respecting the same reaction, they are in equilibrium with each other."

(Nernst, *Theoretical Chemistry*, p. 672. See also pp. 132, 137, 495 for applications of law.)

ACCELERATION OF CHEMICAL REACTIONS WITH ELEVATION OF TEMPERATURE.

It is a general principle of *chemical kinetics* that the velocity with which a chemical system proceeds toward its state of equilibrium increases very greatly with increase in temperature.

(Nernst, *Theoretical Chemistry*, pp. 679, 680. Also Walker, *Introduction to Physical Chemistry*, p. 300 *et seq.*)

HEATS OF REACTION: LAW OF CONSTANT HEAT-SUMMATION.

In processes which occur in nature the associated energy changes may be discriminated as:

1. Production or absorption of heat.
2. Performance of external work.
3. Variation of the internal energy of a system.

In a *chemical system* the sum of the heat produced in a reaction and the external work performed is called the "heat of reaction." This heat of reaction (also heat of formation) may be either positive or negative. It represents the change in the total energy of the chemical system.

The total heat generated in a chemical reaction is entirely independent of the steps followed in passing from initial to final state of the system, and this principle—"the law of constant heat-summation"—makes it possible to calculate heats of formation for steps which are chemically impracticable.

(Nernst, *Theoretical Chemistry,* pp. 592, 597 *et seq.* Also *Smithsonian Physical Tables,* p. 212.)

HEAT OF FORMATION.

By "heat of formation of a chemical compound is meant the quantity of heat which is given off in the formation of the compound from its component elements." The sum of the heats of formation of the substances formed by chemical reaction minus the sum of the heats of formation of the substances used up, is equal to the heat of reaction.

(Nernst, *Theoretical Chemistry,* p. 606. See also 607, 608, where it is shown how *heats of formation* may be obtained from *heats of combustion.*)

CATALYSIS.

Many chemical reactions are observed to take place at an accelerated rate when they occur in the presence of certain substances which themselves suffer no chemical change. Berzelius gave to this phenomenon the name *catalysis.* The name means, " an increase in velocity of reaction caused by the presence of substances which do not take part in it (or only to a secondary extent) although the reaction is capable of taking place without their presence."

A substance which produces catalysis is called a *catalyser.*

(Nernst, *Theoretical Chemistry,* p. 581. Also Washburn, *Principles of Physical Chemistry,* pp. 274, 275.)

CRYSTALLOIDS AND COLLOIDS.

Investigations of the phenomena of diffusion show that substances can be divided into two classes, "crystalloids" and "colloids." The former diffuse more rapidly and as a rule are obtained in crystalline form while the latter are amorphous. Graham named the process for separating the two classes by means of an animal membrane, "dialysis."

(Consult Walker, *Introduction to Physical Chemistry,* p. 233. Also *Ganot's Physics, art.* 140.)

ABSORPTION OF RADIANT HEAT.

It is a general principle that bodies absorb radiant heat which proceeds from heated bodies of the same kind. Also, "any substance is particularly transparent to radiation which has already been sifted by a plate of that substance."

(Preston, *Theory of Heat,* p. 549.)

ABSOLUTE EMISSIVE POWER (Definition).

The *absolute emissive power* of a body for a particular wave-length is the energy at that wave-length radiated per second by unit surface at temperature 1° absolute to surrounding enclosure at absolute zero.

(Preston, *Theory of Heat,* p. 588.)

MONOCHROMATIC EMISSIVE POWER.

The *monochromatic emissive power* of a body is the ratio of the energy, having wave-lengths lying between λ and $\lambda + d\lambda$, which it radiates at absolute temperature T to the energy of the same wave-lengths which a perfectly black body radiates at the same temperature and under exactly similar circumstances.

Call J_λ' the energy of these wave-lengths radiated by a unit surface of the body in the unit of time and call $J\lambda$ the energy radiated by a black-body under the same circumstances. Then $\frac{J_\lambda'}{J_\lambda} = e_\lambda$, the monochromatic emissive power of the body. It is in general a function of the absolute temperature T, the wave-length λ and varies with the nature of the body.

(Consult Preston, *Theory of Heat,* p. 587. Also Chwolson, *Traité de Physique,* Vol. II, Part 8, pp. 55, 56. For experimental methods and values of emissivity, see *Bulletin of the Bureau of Standards,* Vol. II, p. 41, p. 591 and p. 607; articles by Burgess, Waltenberg and Foote.)

ABSORPTIVE POWER AND DEFINITION OF A BLACK-BODY.

Let radiation of a given wave-length λ and of energy E per unit volume fall upon a body. Denote this radiant energy by E_λ. Then if a portion E'_λ of this radiant energy is absorbed by the body, the *absorptive power* of the body for wave-length λ is,

$$a_\lambda = \frac{E'_\lambda}{E_\lambda}$$

If this fraction is unity for all values of λ the body is called, according to Kirchhoff, a perfectly "black-body."

No body has been found which is perfectly black as Kirchhoff defines black. If, however, any body is placed in an enclosure the walls of which are at a uniform temperature, the body will finally assume the temperature of the enclosure and when it does it will emit in quantity and quality (*i.e.,* wave-length) as much radiation as it receives. It will then be indistinguishable to the eye from the neighboring bodies and is said to be at "black-body" temperature.

(Consult Burgess and LeChatelier, *Measurement of High Temperatures,* Chap. VI.)

KIRCHHOFF'S LAW REGARDING RADIATION.

For radiations of the same wave-length and the same temperature, the ratio of the *emissive* and *absorptive powers* is the same for all bodies and is equal to the emissive power of a perfectly *black-body.* In symbols,

$$C_\lambda = \frac{e_\lambda}{a_\lambda},$$

where e_λ is the emissive power of the body for wave-length λ, and a_λ is its absorptive power for wave-length λ, and C_λ is the emissive power of a black-body.

(Preston, *Theory of Heat,* p. 588. Also Burgess and

LeChatelier, *Measurement of High Temperatures*, pp. 243–245. Also Chwolson, *Traité de Physique*, Vol. II, Part 8, pp. 57–59.)

PROPOSITIONS DEDUCIBLE FROM KIRCHHOFF'S RADIATION LAW.

1. The emissive power of an absolutely black-body is the greatest emissive power possible.
2. Every body absorbs the rays which it emits at a given temperature—but every body does not necessarily emit all rays which it absorbs at a given temperature.
3. Every body can absorb rays which it emits at a given temperature, and it can also absorb other rays, provided that among these latter there are rays which a black-body emits at the given temperature.
4. Every body which emits, at a given temperature and under given conditions in a particular direction (under a given angle with the normal), rays of wave-length λ and of a definite type of vibration (character of polarization), absorbs, at the same temperature and under the same conditions, rays of the same wave-length and the same type of vibration which fall upon it in the same direction.
5. The ratio of the emissive to the absorptive power, which is the same for all bodies for the same given values of temperature and wave-length, does not depend upon the kind of vibration; namely, upon the character of the rays emitted and absorbed, in respect to their polarization.
6. The law of Kirchhoff applies to any composite flux of calorific energy (where the wave-lengths are comprised between any arbitrary limits λ_1 and λ_2) if the integral absorption be referred to a flux, which has for its source an absolutely black body taken at the same temperature as the bodies which are to be intercompared.

7. Kirchhoff's law holds true for a composite flux when two given bodies are at the same temperature and when each of them acts as source of the flux, the integral absorption of which, by the other body, is measured.
8. In a closed space all parts of which are at the same temperature all bodies inside and the walls of the enclosure itself produce definite radiation which is identical with the radiation from an absolutely black body.

(For full explanation and proof of the above propositions, consult Chwolson, *Traité de Physique,* Vol. II, Part 8, pp. 59–70.)

INTENSITY OF RADIANT HEAT.

The quantity of heat proceeding from a point-source of heat which is received on a unit surface in the unit of time may be called the intensity of radiant heat. It varies with the temperature of the source and is inversely proportional to the square of the distance from the source.

(*Ganot's Physics, art.* 420.)

HEAT-RADIATION AT AN OBLIQUE ANGLE.

The intensity of oblique rays of radiant heat is proportional to the cosine of the angle which these rays form with the normal to the surface. This " law of the cosine " is not, however, general.

Radiant heat is only one section of the spectrum which extends from wave-lengths shorter than those of ultraviolet light to the longest waves observed. Hence, the laws of reflection and refraction are the same for rays of radiant heat as they are for light.

(Fourier, *The Analytical Theory of Heat,* p. 34. Also *Ganot's Physics, art.* 420.)

STEFAN-BOLTZMANN RADIATION LAW.

The energy radiated in unit time by a black-body is proportional to the fourth power of the absolute temperature, or

$$E = K (T^4 - T_0^4),$$

where E is the total energy radiated by the body at absolute temperature T to the walls of an enclosure at absolute temperature T_0 and K is a constant.

(Preston, *Theory of Heat*, p. 590, and pp. 596–598. Also *Bulletin of the Bureau of Standards*, Vol. I, p. 198. Also Chwolson, *Traité de Physique*, Vol. II, Part 8, p. 71 *et seq.*)

PRESSURE OF RADIATION.

When radiant energy is incident perpendicularly on a plane-surface which is absolutely black, it exerts a pressure on the surface equal to the density of the energy of radiation. If the body is perfectly reflecting the pressure is twice as great.

(Maxwell, *Electricity and Magnetism*, Vol. II, *art.* 792. Also Chwolson, *Traité de Physique*, Vol. II, Part 8, p. 84.)

WIEN'S DISPLACEMENT LAW (1st statement).

"If radiation of a particular wave-length corresponding to a definite temperature is adiabatically altered to another wave-length, then the temperature changes in the inverse ratio." Or,

$$T = \frac{\text{a constant}}{\lambda}$$

(Consult Preston, *Theory of Heat*, p. 600. Also p. 602 for confirmatory experiments.)

WIEN'S DISPLACEMENT LAW (2nd statement).

"When the temperature increases, the wave-length of every monochromatic radiation diminishes in such a way that the product of the temperature and the wave-length is a constant." Or,

$$\lambda T = \lambda_0 T_0.$$

Hence for the wave-length of maximum energy

$$\lambda_m T = \text{a constant.}$$

(*Bulletin of the Bureau of Standards,* Vol. I, No. 2, p. 202.)

WIEN'S LAW OF RADIATION OF WAVE-LENGTH OF MAXIMUM ENERGY.

The energy radiated from a black-body source which corresponds to the wave-length having a maximum energy is proportional to the fifth power of the absolute temperature. Or,

$$E_{max.} = BT^5,$$

where T is the absolute temperature and B is a constant.

(*Bulletin of the Bureau of Standards,* Vol. I, p. 202. Also Preston, *Theory of Heat,* pp. 601, 610. Also Chwolson, *Traité de Physique,* Vol. II, part 8, p. 73.)

WIEN'S LAW OF SPECTRAL DISTRIBUTION OF ENERGY.

$$J = c_1 \lambda^{-5} e^{-\frac{c_2}{\lambda T}}$$

where J = the energy corresponding to wave-length λ,
T = absolute temperature of the radiating black-body,
e = base of the natural system of logarithms and
c_1 and c_2 are constants.

(*Bulletin of the Bureau of Standards,* Vol. I, p. 204. For use of the equation in pyrometry, see p. 210.)

PLANCK'S LAW OF SPECTRAL DISTRIBUTION OF ENERGY.

$$J = c_1 \lambda^{-5} \frac{1}{e^{\frac{c_2}{\lambda T}} - 1}$$

where J is the energv corresponding to wave-length λ, T is the absolute temperature, e the base of the natural system of logarithms and c_1 and c_2 are constants.

Planck's law agrees with experiment better than Wien's and holds well for a wide variation in λ.

(*Bulletin of the Bureau of Standards,* Vol. I, p. 206.)

PREVOST'S THEORY OF EXCHANGES.

If two bodies are associated in such a manner that one receives the radiation of the other, each radiates independently and the temperature of either will fall if it radiates more energy than it absorbs.

(*Ganot's Physics, art.* 421.)

THE QUANTUM HYPOTHESIS.

The only hypothesis which has satisfactorily accounted for the laws of radiation from black bodies and many phenomena, such as the emission of electrons from illuminated metals, has been based on the supposition that radiant energy is absorbed or emitted or both not continuously, but by discontinuous, discrete units of magnitude proportional to the frequency of the radiation.

(Planck, *Vorlesungen über die Theorie der Wärmestrahlung.*)

LAW OF PHOTO-CHEMICAL REACTION.

Abundant research has led to the result that when a photo-chemical system is illuminated the resultant action depends both on the light-intensity and the time the illumination continues. Hence the law: "When light of the same kind is used, the photo-chemical action depends solely on the product of the intensity and the duration of the exposure." This law applies in photography and for X-Ray exposures.

(Nernst, *Theoretical Chemistry,* p. 786.)

CRYSTALS: THE LAW OF INTERFACIAL ANGLES.

A crystal is a homogeneous body the various physical properties of which are differently manifested when considered along different lines radiating from any point in the body.

The fundamental law of crystallography states that, "the inclination of two definite crystal planes to each other, for the same substance, and measured at the same temperature, is constant and independent of the size and development of the planes." This law is not, however, quite rigid.

(Nernst, *Theoretical Chemistry,* pp. 72, 73. Also Spencer, *The World's Minerals,* p. 15.)

NEUMANN'S LAW OF CRYSTAL-ZONES.

A set of planes of a crystal which intersect in such manner that the lines of intersection are parallel to one another is called a *zone* and this common direction is called the "zonal axis." Otherwise stated a crystal has a girdle of faces (a zone) the edges of which form parallel lines.

It is a fundamental law of crystallography that all planes which can occur on a crystal are related to each other in zones or from any four planes, no three of which lie in any one zone, all crystal planes can be derived by means of zones.

(Nernst, *Theoretical Chemistry,* p. 73. Also Spencer, *The World's Minerals,* pp. 17, 18.)

V

ELECTRICITY AND MAGNETISM

ELECTRICITY AND MAGNETISM

ELECTRIFICATION PRODUCED BY FRICTION.

If two unlike substances which are insulators are rubbed together and then separated they are found to be *electrified* with equal quantities of electricity, the one substance with positive, or vitreous, and the other with negative, or resinous, electricity.

(Maxwell, *Treatise on Electricity and Magnetism,* Vol. I, *art.* 27.)

ELECTRIFICATION PRODUCED BY INDUCTION.

When an electrified body is suspended in a hollow conducting vessel without touching it, the outside of the vessel acquires electrification of the same sign as the electrified body, and the inside of the vessel acquires electrification of the opposite sign. The electrification of the vessel thus produced is called electrification by *induction.*

(Maxwell, *Treatise on Electricity and Magnetism,* Vol. I, *art.* 28. Also *Ganot's Physics, art.* 764.)

ELECTRIFICATION BY CONDUCTION.

When an originally unelectrified and insulated metal body is connected with an electrified body by means of a conducting wire, which is itself insulated, the first body becomes electrified by a passage of electricity over the wire. This passage is called electrification by *conduction.*

(Maxwell, *Treatise on Electricity and Magnetism,* Vol. I, *art.* 29.)

CONDUCTORS AND INSULATORS.

At low and at ordinary temperatures materials are separable into two fairly distinct classes: those which readily conduct electricity, called *conductors,* and those which conduct electricity very slightly, or not at all, called *insulators.* Above 1200° or 1500° C. the distinction between the two classes disappears rapidly as the temperature increases.

(See "Methods, Data and New Apparatus for Measuring Electrical Conductivity above 1500° C. of Vapors at Normal Pressure," by E. F. Northrup, *Jour. of the Franklin Institute,* March, 1915. Also *Ganot's Physics, art.* 743.)

EQUALITY OF POSITIVE AND NEGATIVE ELECTRICITY.

When electrification is excited by any means the quantities of positive and negative electricity produced, or revealed, are always equal.

Modern theory asserts that the ultimate unit of negative electricity is a definite quantity of electricity called an *electron.* Millikan concludes from experiments that the charge carried by an electron is 4.774×10^{-10} electrostatic unit. The ratio, $\frac{e}{m}$, of the charge (electrostatic units) to the mass (grams) of an electron is 5.31×10^{17}, provided the velocity of the electron is a small fraction of the velocity of light.

(Thomson, *Elements of Electricity and Magnetism,* p. 10, See also Campbell, *Modern Electrical Theory,* pp. 25–28, 78–80. Also *Smithsonian Physical Tables,* p. 342.)

LAW OF REPULSION OF ELECTRIC CHARGES.

When two charged bodies are at a distance r apart, r being very large compared with the greatest linear dimensions of either of the bodies, the repulsion between them is *proportional* to the product of their charges and inversely proportional to the square of the distance between them.

The repulsion between two charges Q_1 and Q_2 in air is, in electrostatic units, $F = \frac{Q_1 Q_2}{r^2}$ dynes.

(Thomson, *Elements of Electricity and Magnetism,* p. 12. Also, for simple proof, see Lommel, *Experimental Physics,* p. 285. Also *Ganot's Physics, art.* 753.)

POTENTIAL.

The work which must be done by an external agent to bring a unit of positive electricity, by any path, from an infinite distance (or from a place where the potential is zero) to a given point in space is called the *potential* at the point.

A body charged positively always tends to move from places of greater to places of less positive potential, and a body charged negatively always tends to move in the opposite direction.

(Maxwell, *Treatise on Electricity and Magnetism,* Vol. I, *arts.* 45, 70.)

ELECTROMOTIVE FORCE.

The difference of potential between two conductors or between two points in space or between two points on a body is equal to the *electromotive force* (e.m.f.) between them. The e.m.f. between the two points in a circuit is equal to the product of the current and the ohmic resistance between the two points.

When a charge e moves along a given path from a point A to a point B and work W is done by the electric force, such that W = Ee, the quantity E is called the total electromotive force acting between the points A and B. In electrostatics E and $V_1 - V_2$, the potential difference between the points, are identical quantities.

(Maxwell, *Treatise on Electricity and Magnetism,* Vol. I, *arts.* 45, 49, 69, 241.)

POTENTIAL DUE TO A SYSTEM OF POINT-CHARGES.

When any number of electrified points having charges e_1, e_2, e_3, etc., are distributed through space, then if r_1, r_2, r_3, etc., are the distances of these points respectively from a point P in space, the potential at P due to the system is,

$$V = \Sigma\left(\frac{e}{r}\right)$$

(Maxwell, *Treatise on Electricity and Magnetism*, Vol. I, *art*. 73.)

FORCE BETWEEN CHARGED BODIES VARIED BY THE MEDIUM.

When the charges are given, the mechanical forces on bodies in an electric field are diminished by the interposition of a medium with a larger specific inductive capacity. Thus the force between two point-charges Q_1 and Q_2, a distance r apart in a medium of specific inductive capacity K, is,

$$F = \frac{Q_1 Q_2}{Kr^2}.$$

(Thomson, *Elements of Electricity and Magnetism*, p. 129.)

ELECTRIC EQUILIBRIUM.

A conductor can only be in electric equilibrium when every point in it is at the same potential. This potential is called the Potential of the Conductor.

(Maxwell, *Treatise on Electricity and Magnetism*, Vol. I, *art*. 45. Also consult, for an extended and clear exposition of the facts and laws of electricity in equilibrium, Chwolson, *Traité de Physique*, Vol. 4, Part 10, Chap. I. See p. 81.)

ELECTRIC ABSORPTION.

The phenomenon of electric absorption is *not* an actual absorption of electricity, for if a condenser is in the interior of a hollow electric conductor there is no alteration in its surface electrification by the "absorption" taking place in the condenser. Or, as stated by Faraday: "It is impossible to charge matter with an absolute and independent charge of one kind of electricity."

(Maxwell, *Treatise on Electricity and Magnetism,* Vol. I, *arts.* 53, 54.)

ELECTRIC INTENSITY INSIDE AN INCLOSED CONDUCTING SURFACE.

There is *no* electric intensity or lines of electric force on the inside of a surface which is conducting and which completely incloses a portion of space, however highly this surface is charged, if there is no charged body in the space.

(Thomson, *Elements of Electricity and Magnetism,* pp. 21, 22, 29. Also *Ganot's Physics, art.* 765.)

GAUSS' THEOREM.

"The total normal electric induction over any closed surface drawn in the electric field is equal to 4π times the total charge of electricity inside the closed surface."

(Thomson, *Elements of Electricity and Magnetism,* p. 14.)

COULOMB'S LAW.

The electric intensity of a point p close to the surface of a conductor surrounded by air is at right angles to the surface, It is equal to $4\pi\ \sigma$ where σ is the surface density of the electrification. If the surface of the conductor is in contact with a dielectric of specific inductive capacity K, then the electric intensity at the point p is,

$$R = \frac{4\pi}{K}\sigma.$$

(Thomson, *Elements of Electricity and Magnetism,* pp. 36, 122.)

ENERGY OF A SYSTEM OF CONDUCTORS.

The potential energy of a system of charged electric conductors placed in an electric field which arises from the charges on the conductors of the system is equal to one-half the sum of the products obtained by multiplying the charge on each conductor by its potential, or,

$$E = \tfrac{1}{2}\Sigma QV,$$

where Q is a charge and V the potential of that charge.

(Thomson, *Elements of Electricity and Magnetism,* p. 37.)

ELECTRICITY AND AN INCOMPRESSIBLE FLUID COMPARED.

"The motions of electricity are like those of an incompressible fluid, so that the total quantity within an imaginary fixed closed-surface remains always the same."

(Maxwell, *Treatise on Electricity and Magnetism,* Vol. I, *art.* 61.)

WORK DONE IN A DISPLACEMENT OF AN ELECTRIFIED SYSTEM.

The work done by the electric forces during the displacement of an electrified system, when the potentials are maintained constant, is equal to the *increment* of the electric energy. The work done, therefore, by a battery which maintains the potentials constant is *twice* the work done by the system during its displacement.

(Maxwell, *Treatise on Electricity and Magnetism,* Vol. I, *art.* 93c.)

POTENTIAL ENERGY-CHANGES.

If any small displacement of a system of electrified conductors takes place the *diminution* in the electric energy of the system, when the *charges* are kept constant, is equal to the *increase* in the potential energy when the same displacement takes place and the *potentials* are kept constant.

(Thomson, *Elements of Electricity and Magnetism,* p. 54.)

MECHANICAL FORCE AT THE SURFACE OF A CHARGED CONDUCTOR.

The surface of every charged conductor is subject to a *mechanical force* which acts outward along the normal. The value of the force per unit area of the surface of the conductor for *any* dielectric surrounding the conductor, is, $F = \frac{1}{2} R\sigma$, and, if the dielectric surrounding the conductor is air,

$$F = \frac{R^2}{8\pi} = 2\pi\,\sigma^2.$$

Here R is the electric intensity and σ is the density of the surface electrification. The maximum value of F in air at normal pressure and 15° C. is about $\frac{10^4}{8\pi}$ dynes per square centimeter, which is a pressure of about 0.3 mm. of mercury.

(Thomson, *Elements of Electricity and Magnetism*, pp. 58, 59.)

PASCHEN'S LAW FOR SPARKING POTENTIALS IN A GAS.

The sparking potential between electrodes in a gas depends on the length of the spark-gap and the pressure of the gas in such a way that it is directly proportional to the mass of gas between the two electrodes. Or we can consider the sparking potential as a function of the pressure × the density of the gas.

(Thomson, *Conduction of Electricity Through Gases*, pp. 451–460.)

STATE OF ELECTRIC STRESS IN A MEDIUM.

When bodies in a dielectric medium are electrified, "at every point of the medium there is a state of stress such that there is tension along the lines of force and pressure in all directions at right angles to these lines, the numerical magnitude of the pressure being equal to that of the tension, and both varying as the square of the resultant force at the point."

(Maxwell, *Treatise on Electricity and Magnetism*, Vol. I, *art.* 109.)

ELECTRIC DISPLACEMENT.

The whole quantity of electricity Q, which is displaced through a unit area, when the electric intensity R is normal to the area, is $Q = \frac{K}{4\pi}R$; where K is the specific inductive capacity of the dielectric.

(Maxwell, *Treatise on Electricity and Magnetism*, Vol. I, *art*. 62. Compare with statement in Thomson, *Elements of Electricity and Magnetism*, p. 122.)

ENERGY OF ELECTRIC FIELD.

The electric field represents a certain store of energy in the medium. If R is the intensity of the field the stored energy *per unit volume* is $E = \frac{K}{8\pi}R^2$, where K is the specific inductive capacity of the medium.

(Thomson, *Elements of Electricity and Magnetism*, pp. 70–72, and 123.)

SPECIFIC INDUCTIVE CAPACITY.

The charge of electricity which a condenser will hold when charged to a given potential is dependent upon the nature of the dielectric. The ratio of the charge which the condenser holds when a given dielectric is used, to the charge it holds when air (or a perfect vacuum) is the dielectric, is called the "specific inductive capacity" of the dielectric. This ratio is either equal to or greater than unity.

(See Thomson, *Elements of Electricity and Magnetism*, Chap. IV. Also *Ganot's Physics, art*. 783.)

CONDUCTORS AND DIELECTRICS IN AN UNUNIFORM ELECTRIC FIELD.

A conductor placed in an ununiform electric field tends to move from the weak to the strong parts of the field; so likewise does a dielectric surrounded by one of smaller specific inductive capacity.

(Thomson, *Elements of Electricity and Magnetism*, p. 128.)

ELECTRIC INTENSITY INSIDE AND OUTSIDE CONDUCTORS AND DIELECTRICS.

The electric intensity inside a conductor placed in an electric field vanishes, and just inside a dielectric of greater specific inductive capacity than the surrounding medium the electric intensity is less than that just outside.

(Thomson, *Elements of Electricity and Magnetism*, p. 128.)

EARNSHAW'S THEOREM ON STABILITY.

If a charged body is placed in an electric field and is altogether free to move, it is always in unstable equilibrium in respect to translational motion.

(For proof, see Maxwell, *Treatise on Electricity and Magnetism*, Vol. I, *art.* 116. Also Jeans, *Electricity and Magnetism*, p. 165.)

CAPACITY OF CONDENSERS, JOINED IN PARALLEL AND IN SERIES.

The total capacity C of a number of condensers joined in *parallel* is equal to the sum of the individual capacities C_1, C_2, C_3, etc., and the reciprocal of the total capacity of a number of condensers, joined in *series*, is equal to the sum of the reciprocals of the individual capacities. Thus for parallel combination

$$C = C_1 + C_2 + C_3 + \text{etc.}$$

and for series combination

$$\frac{1}{C} = \frac{1}{C_1} + \frac{1}{C_2} + \frac{1}{C_3} + \text{etc.}$$

(Thomson, *Elements of Electricity and Magnetism*, pp. 110–113. Also *Ganot's Physics*, *art.* 799.)

MAGNETISM (Definitions).

1. The *magnetic moment* of a magnet is, $M = ml$ where m is the strength of its pole and l the distance between its poles.
2. The *intensity of magnetization* of a magnetizable substance, uniformly magnetized, is $I = \frac{M}{V}$, where V is its volume and M is its magnetic moment.

(Consult Jeans, *Electricity and Magnetism*, pp. 355, 358.)

3. A *magnetic shell* is a thin sheet of magnetizable substance magnetized at each point in the direction of the normal to the sheet at that point.
4. The *strength of a magnetic shell* is the intensity I of its magnetization times t its thickness. Thus,

$$\vartheta = It = \frac{M}{S} = \text{magnetic moment per unit area.}$$

5. The *magnetic potential* at a point due to a magnetic shell of uniform strength is $V = \theta\omega$, where ω is the solid angle at the point subtended by the contour of the shell.

(Consult Jeans, *Electricity and Magnetism*, pp. 365, 366.)

FORCE ACTING BETWEEN MAGNETIC POLES.

The force between two magnetic poles in air is in the straight line joining them, and is numerically equal to the product of the strengths of the poles divided by the square of the distance between them.

$$\text{Thus, } F = \frac{m\,m'}{r^2},$$

where m and m′ are the strengths of the poles and r is the distance between them. If m and m′ are of unlike sign the force is an attraction, if of like sign the force is a repulsion.

The law as here stated assumes that the strength of each pole is measured in terms of a unit, the magnitude of which is deduced from the terms of the law.

(Maxwell, *Treatise on Electricity and Magnetism,* Vol. II, *arts.* 373, 374. For Gauss' proof of law, see Thomson, *Elements of Electricity and Magnetism,* p. 206. Also *Ganot's Physics, arts.* 716–720.)

POTENTIAL DUE TO A MAGNETIC SOLENOID.

A "Magnetic Solenoid" is a filament of magnetic matter so magnetized that its strength is the same at every transverse section. The Magnetic Potential due to a magnetic solenoid depends only on its strength and the position of its ends, called its poles.

$$\text{Thus, } V = m\left(\frac{1}{r_1} - \frac{1}{r_2}\right),$$

where m is the strength of its poles and r_1 and r_2 are distances from the positive and negative poles respectively to the point where the potential is V.

(Maxwell, *Treatise on Electricity and Magnetism,* Vol. II, *art.* 407.)

TOTAL CHARGE OF MAGNETISM.

The total charge of magnetism reckoned algebraically on any magnet is zero.

(Thomson, *Elements of Electricity and Magnetism,* p. 190. Also Maxwell, *Treatise on Electricity and Magnetism,* Vol. II, *art.* 377. Also *Ganot's Physics, art.* 697.)

MAGNETIC FORCE DUE TO A MAGNET.

If two points P and Q be taken equidistant from the center of a bar-magnet, the point P being in the line of its axis and the point Q in the equitorial plane at right angles to the axis of the magnet, then the magnetic force at P is *twice* the magnetic force at Q. Thus, if OP is the distance to P and OQ is the distance to Q from the center of the magnet,

$$H_P = \frac{2M}{OP^3} \text{ and } H_Q = \frac{M}{OQ^3}$$

where M is the magnetic moment of the magnet.

(Thomson, *Elements of Electricity and Magnetism*, pp. 195–197.)

MAGNETIC FIELDS DUE TO A MAGNETIZED SPHERE AND A SMALL MAGNET COMPARED.

A "uniformly magnetized sphere produces the same effect outside the sphere as a very small magnet placed at its center, the axis of the small magnet being parallel to the direction of magnetization of the sphere, while the moment of the magnet is equal to the intensity of magnetization multiplied by the volume of the sphere."

(Thomson, *Elements of Electricity and Magnetism*, p. 224.)

MAGNETISM INDUCED BY A MAGNETIC FIELD.

All substances which are measurably diamagnetic or paramagnetic when placed in a magnetic field become charged with magnetism, and the *diamagnetic* substances tend to move toward *weaker* portions of the field and the *paramagnetic* substances toward *stronger* portions of the field.

(Thomson, *Elements of Electricity and Magnetism*, p. 242. Also *Ganot's Physics, arts.* 700, 701.)

MAGNETIC INDUCTION.

Magnetic induction is a vector quantity and is the sum of two vectors, the magnetic force H and 4π times the magnetization I.

Thus, $B = H + 4\pi I$.

Tubes of magnetic induction are always continuous and closed tubes.

(Maxwell, *Treatise on Electricity and Magnetism,* Vol. II, *art.* 400. Also Thomson, *Elements of Electricity and Magnetism,* pp. 244–247. Also Jeans, *Electricity and Magnetism,* p. 373.)

SOME RELATIONS OF MAGNETIC QUANTITIES.

Let H = magnetic force,

B = magnetic induction,

I = magnetization,

μ = magnetic permeability and

k = magnetic susceptibility, then

$$B = H + 4\pi I = (1 + 4\pi k)H = \mu H,$$

$$I = \frac{B - H}{4\pi} = \left(\frac{\mu - 1}{4\pi}\right)H = kH,$$

$$H = B - 4\pi I = \frac{B}{1 + 4\pi k} = \frac{B}{\mu} \text{ and}$$

$$\mu = 1 + 4\pi k.$$

(Thomson, *Elements of Electricity and Magnetism,* p. 247. Also Maxwell, *Treatise on Electricity and Magnetism,* Vol. II, *art.* 438.)

MAGNETIC HYSTERESIS.

Magnetic hysteresis is a phenomenon which results from the transformation of magnetic energy into heat when the magnetization of a substance is changed. It is a species of molecular friction and is always exhibited when cyclical reversals of magnetic flux in a magnetizable substance are produced. It is analogous with friction in mechanics.

C. P. Steinmetz gives for the loss of energy due to hysteresis in ergs per cycle per cm^3,

$$W = \eta B^{1.6},$$

where B is the maximum magnetic induction per cm^2 and η is the "coefficient of hysteresis." Steinmetz gives $\eta = 0.0025$ as a fair average value for selected steel.

(Steinmetz, *Alternating Current Phenomena*, Chap. X. See p. 116.)

MAGNETIC AND ELECTRIC ANALOGUES.

There is a complete analogy between the disturbance in the distribution of an electric field produced by the presence of uncharged dielectrics and the disturbance of a magnetic field produced by paramagnetic or diamagnetic bodies in which the magnetism is entirely induced. In this analogy the magnetic force H is equivalent to the electric intensity R and the magnetic permeability μ is equivalent to the specific inductive capacity K.

(Thomson, *Elements of Electricity and Magnetism*, p. 258.)

ELECTROMOTIVE-FORCE SERIES.

Two unlike metals when immersed in an acid or salt solution have acting between them an electromotive force. The potential-difference between the two metals when immersed in a given solution varies with the nature of the metals.

Metals may be arranged in an *electromotive-force series*

in which the most electropositive metal begins and the most electronegative metal ends the series.

(*Ganot's Physics, art.* 817. See also *Smithsonian Physical Tables,* p. 267.)

GALVANIC POLARIZATION.

When a chemical system, originally in equilibrium, is electrolysed the decomposition caused by the electrolysis produces a displacement of equilibrium, and it follows *necessarily* that the current passed through the system has to overcome an *opposing* electromotive force. The development of this opposing electromotive force in a galvanic cell is termed "polarization."

(Nernst, *Theoretical Chemistry,* p. 739. Also Walker, *Introduction to Physical Chemistry,* pp. 374, 375.)

CONSERVATION OF ENERGY IN ELECTROLYSIS: THOMSON'S THEOREM.

"The electromotive force of an electrochemical apparatus is in absolute measure equal to the mechanical equivalent of the chemical action on one electrochemical equivalent of the substance."

(Maxwell, *Treatise on Electricity and Magnetism,* Vol. I, *arts.* 262, 263. For a scientific treatment of the principles of electrochemistry and an analysis of the above statement, consult Nernst, *Theoretical Chemistry,* Chap. VII, pp. 731 *et seq.*)

ELECTROLYSIS LITTLE AFFECTED BY PRESSURE.

If the products of an electrolysis are gases which obey Boyle's law, the product of their pressure and volume will be constant at a given temperature and the e.m.f. required for electrolysis is nearly independent of the pressure. Electrolysis of dilute acids, therefore, cannot be checked by confining the gases of decomposition in a small space.

(Maxwell, *Treatise on Electricity and Magnetism, art.* 263.)

FARADAY'S FIRST LAW OF ELECTROLYSIS.

"The quantity of an electrolyte decomposed by the passage of a current of electricity is directly proportional to the quantity of electricity which passes through it."

(Thomson, *Elements of Electricity and Magnetism*, p. 282. Also *Ganot's Physics, art.* 939.)

FARADAY'S SECOND LAW OF ELECTROLYSIS.

"If the same quantity of electricity passes through different electrolytes the weights of the different ions deposited will be proportional to the chemical equivalents of the ions."

(Thomson, *Elements of Electricity and Magnetism*, p. 282. For a full account and a bibliography, consult Chwolson, *Traité de Physique,* Vol. IV, Part 10, Chap. V. See p. 617 for statements of the two laws of electrolysis.)

A PRINCIPLE OF ELECTROLYTIC DECOMPOSITION.

An electrolytic decomposition can only proceed when the loss of energy in the battery which supplies current to an electrolytic cell is greater than the gain of energy in the electrolyte of the electrolytic cell. An action contrary to the above would violate the principle of the conservation of energy.

(Thomson, *Elements of Electricity and Magnetism*, p. 301.)

THE "FARADAY."

The *Faraday* is the name now commonly used to denote the "quantity of electricity associated with a chemical equivalent in any electrochemical change."

The deposition of silver from a solution of silver-nitrate has been most extensively investigated. G. W. Vinal and S. J. Bates writing in the *Bulletin of the Bureau of Standards,* Jan. 2, 1914, give as the value of the Faraday for silver,

$$\frac{107.88}{0.00111800} = 96494 \text{ international Coulombs.}$$

(Nernst, *Theoretical Chemistry*, pp. 727, 728.)

OHM'S LAW.

The current which flows in a metallic conductor or in an electrolyte is proportional to the difference of potential at the extremities of the conductor. Thus, if E = e.m.f., R = resistance and I = current, $I = \frac{E}{R}$. The quantity R has been experimentally proved to be independent of the strength of the current (the temperature being the same) to at least 1 part in 100,000.

(Consult Maxwell, *Treatise on Electricity and Magnetism*, Vol. I, *art.* 241. Also Thomson, *Elements of Electricity and Magnetism*, p. 284. Also Northrup, *Methods of Measuring Electrical Resistance, art.* 105, p. 16.)

RESISTANCES IN SERIES.

When several resistances are joined in series the total resistance is equal to the sum of the individual resistances.

From coils which have resistance-values 1–1–4–3 or 1–3–3–2 (or any multiple of these values) the successive values 0 to 9 can be obtained by moving a single plug-connector.

(Northrup, *Methods of Measuring Electrical Resistance, art.* 503, p. 82.)

RESISTANCE OF A NUMBER OF CONDUCTORS ARRANGED IN PARALLEL.

When a number of resistances, R_1, R_2, R_3, etc., are joined together in parallel combination the reciprocal of the total resistance R is equal to the sum of the reciprocals of the individual resistances.

$$\text{Thus,}\quad \frac{1}{R} = \frac{1}{R_1} + \frac{1}{R_2} + \frac{1}{R_3} + \ldots\ldots$$

Or, calling the reciprocal of any resistance a *conductance*, the total conductance = the sum of the conductances of individual conductors when joined in parallel combination.

If n resistance-units are used singly, and joined in all possible *series* combinations, $2^n - 1$ resistance-combinations can be obtained. The total number N of combinations possible for n units used *singly* and joined in *series* and *parallel* combinations is,

$$N = 2^{n+1} - (n+2).$$

(*Ganot's Physics, art.* 853. Also Northrup, *Methods of Measuring Electrical Resistance, art.* 501, p. 80.)

CONDITION FOR A DEFINITE RESISTANCE.

The conductor must be considered as having its surface divided into *three* portions:

1. a portion over which the potential is maintained constant,
2. a portion over which the potential is held constant but higher or lower than the first and
3. a remaining portion, which is impervious to electricity.

Only when the conductor is in approximately the above condition can its resistance be said to be definite.

(Maxwell, *Treatise on Electricity and Magnetism,* Vol. I, *art.* 306.)

RESISTANCE OF A CONDUCTOR CHANGES WHEN SPECIFIC RESISTANCE CHANGES.

If the specific resistance of any portion of a conductor is increased, that of the remainder being unchanged, the resistance of the whole conductor will be increased, and if the specific resistance of any portion of it is decreased, that of the remainder being unchanged, the resistance of the whole conductor will be decreased.

(Maxwell, *Treatise on Electricity and Magnetism,* Vol. I, *art.* 306.)

FOUR-TERMINAL CONDUCTORS.

"The generalized four-terminal conductor is a mass of conducting material of any size or shape and has four limited portions of the surface arbitrarily selected and adapted for making electrical connection to other conductors."

This definition (given by Frank Wenner, *Bulletin of the Bureau of Standards,* Vol. 8, p. 560) covers those standards of resistance having two current and two potential terminals, which are used in fall of potential methods and with the Kelvin-double bridge, in the measurement of resistance.

KIRCHHOFF'S THEOREM ON INTERCHANGE OF ELECTRODES.

"In any conductor or system of conductors having four terminals, 1, 2, 3, and 4 selected in any way, the drop in potential from 1 to 2 caused by a current entering at 3 and leaving at 4, is equal to the drop in potential from 3 to 4 caused by an equal current entering at 1 and leaving at 2."

This theorem, given by Kirchhoff in 1847, is of importance in connection with the measurement of resistances by fall of potential methods.

(See *Bulletin of the Bureau of Standards,* Vol. VIII, p. 563.)

JOULE'S LAW OF GENERATION OF HEAT IN A CONDUCTOR.

The heat produced by the passage of an electric current through a solid metallic conductor is proportional to the product of the resistance of the conductor, the square of the current and the time, or to the product of the applied e.m.f. the current and the time.

Thus, $JH = RI^2t = EIt$,

where J is Joule's dynamical equivalent of heat, H the number of units of heat, R the resistance of the conductor, I the current, t the time during which the current flows and E the applied e.m.f. When practical units are used, $H = 0.24\ RI^2t$ in gram-calories, approximately.

(Consult Maxwell, *Treatise on Electricity and Magnetism,* Vol. I, *art.* 242. Also "High Temperature Investigation and a Study of Metallic Conduction," by E. F. Northrup, *Jour. of the Franklin Institute,* June, 1915, pp. 650–652.)

CONVERSION OF MECHANICAL ENERGY INTO HEAT IN A CONDUCTOR.

The mechanical work done by electromotive force in driving electricity through a solid metallic conductor is entirely converted into heat. If, however, the metallic conductor is liquid (molten) some power is spent in circulating the liquid metal.

(Consult paper by E. F. Northrup, "A New Type of Ammeter, etc." *Proc. of the American Electrochemical Society,* May 7, 1909, pp. 303–329.)

MINIMUM HEAT CONDITION.

"In any system of conductors in which there are no internal electromotive forces the heat generated by currents distributed in accordance with Ohm's law is less than if the currents had been distributed in any other manner consistent with the actual conditions of supply and outflow of the current."

(Maxwell, *Treatise on Electricity and Magnetism,* Vol. I, *art.* 284. Also Thomson, *Elements of Electricity and Magnetism,* pp. 310–313.)

THE WIEDEMANN-FRANZ RATIO.

The best conductors of electricity are also the best conductors of heat. The ratio of the thermal to the electrical conductivity for all good conductors at the same temperature has the same value; the value of this ratio is proportional to the absolute temperature. This law holds well for pure metals, less accurately for alloys and not at all for poor conductors.

(Campbell, *Modern Electrical Theory,* pp. 66–68. For experimental illustrations see Richardson, *The Electron Theory of Matter,* pp. 410–413.)

SUPERCONDUCTIVITY.

When the temperature of a very pure metal is reduced to near the absolute zero of temperature (to less than 4° K.) its electrical resistance vanishes or its conductivity becomes practically infinite.

This is an experimental result of the researches of Kamerlingh Onnes and has been termed "Superconductivity."

(Consult, "Electrical Conductivity at High Temperatures and Its Measurement," by E. F. Northrup, *Trans. of the Amer. Electrochem. Soc.,* Vol. XXV, 1914, p. 377. For data see *Smithsonian Physical Tables,* p. 280.)

RESISTANCE-TEMPERATURE RELATIONS FOR METALS.

The variation in electrical resistance per unit of resistance per degree is called the "resistance-temperature coefficient" of a substance. If R_t is the ohmic resistance of a sample of the substance when at temperature t then its resistance-temperature coefficient at this temperature is

$$\alpha_t = \frac{1}{R_t} \frac{dR_t}{dt}.$$

A few general relations between resistance and temperature for several pure metals and some alloys have been obtained experimentally. Thus:

1. All pure metals when in the solid state increase in resistance when the temperature is increased and the coefficient α_t is approximately the same for all pure metals and has the same order of magnitude as the coefficients of expansion of gases.

(Northrup, *Jour. of the Franklin Institute,* June, 1915, p. 636 *et seq.*)

2. Several pure metals (Ag, Au, Cu, Pb, Al, Sn, Sb, Bi) when in the molten state increases in resistance linearly with the temperature, namely, the coefficient α_t is strictly a constant over a considerable range of temperature. The same is true for some alloys; Sn + Bi in particular.

(Northrup, *Jour. of the Franklin Institute,* June, 1915, Figs. 1 and 2, pp. 638, 639.)

3. If we call $\beta_t = \frac{1}{V_t}\frac{dV_t}{dt}$ the coefficient of volume-expansion of a molten metal, then the ratio $\frac{\alpha_t}{\beta_t}$ is nearly the same quantity (lying within the limits 3.48 and 5.19) for at least six pure metals (Na, K, Sn, Hg, Pb, Bi).

(Northrup, *Trans. of the Amer. Electrochem. Soc.,* Vol. XXV, 1914, p. 388)

4. The resistance of most metals (Na, Al, K, Cu, Zn, Cd, Sn, Sb, Bi, Au, Hg, Pb, have been studied) upon changing from the solid to the molten state approximately doubles. Antimony and bismuth are exceptions to the general rule, the resistance of these decreasing when fusion occurs.

(See reference under 3)

5. The change in the resistivity per degree C. of a sample of copper is 0.00681 microhm per centimeter cube, or the conductivity of copper is strictly proportional, over ordinary ranges of temperature (as 10° to 100°C.), to its resistance-temperature coefficient. Thus either may be deduced from knowledge of the other.

(J. H. Dellinger, "The Temperature Coefficient of Resistance of Copper," *Bulletin of the Bureau of Standards,* Vol. 7, pp. 83, 84.)

6. When the resistivity is given, at any temperature, for pure tin and pure bismuth—both being in the molten state—then the resistivity of any molten alloy of known proportions of these two metals may be calculated, because the resistivity of the alloy bears a strictly linear relation to the percentage of gram-atoms in which either constituent is present.

(Northrup and R. G. Sherwood, "New Methods for Measuring Resistivity of Molten Materials; Results for Certain Alloys," *Jour. of the Franklin Institute,* Aug., 1916.)

7. At temperatures which exceed about 1452° C., the melting-point of nickel, all known substances in either the solid or liquid state are more or less electrically conducting and above this temperature it is impossible by any means to obtain, even approximately, good electrical insulation.

(Northrup, *Jour. of the Franklin Institute,* March, 1915, p. 352.)

A GENERAL RELATION BETWEEN CONDUCTANCE AND CAPACITY.

Let two perfect conductors which serve as electrodes be immersed in an electrically conducting homogeneous medium of electrical conductivity σ and call G the electrical conductance between the two electrodes. Then substitute for the electrically conducting medium a dielectric medium of specific inductive capacity K and call C the electrostatic capacity between the same two electrodes. Under these circumstances the general relation holds

$$\frac{G}{C} = \frac{4\pi\sigma}{K}.$$

(Northrup, " Use of Analogy in Viewing Physical Phenomena," *Jour. of the Franklin Institute,* July, 1908, pp. 31, 32. Also consult Jeans, *Electricity and Magnetism,* pp. 339, 340.)

A RELATION BETWEEN CAPACITY AND RESISTANCE.

In every portion of an electric circuit where electrostatic lines and lines of current-flow pass through the same medium, the product of the resistance and the capacity, however dimensions are varied, is constant and equal to $\frac{\rho K}{4\pi}$. Here ρ is the resistivity of the medium and K its specific inductive capacity. This proposition is of importance in the electrical measurement of high resistances.

(Northrup, *Methods of Measuring Electrical Resistance, art.* 902, pp. 186–189.)

ELECTROMOTIVE FORCES IN SERIES.

Electromotive forces add in series as scalar quantities but obey the law of algebraic signs.

Thus, $E = e_1 + e_2 + e_3 + (-e_4) + (-e_5) + e_6 + - - - - -$.

(Consult Jeans, *Electricity and Magnetism,* pp. 298, 299.)

CONTACT ELECTRICITY.

When two different metals are in contact there is in general an e.m.f. acting from one to the other. Let C be taken as a standard metal: then if the potential of a metal I in contact with C at zero potential is i and that of a metal Z in contact with C at zero potential is z, the potential of Z in contact with I at zero potential is z – i.

(Maxwell, *Treatise on Electricity and Magnetism,* Vol. I, *arts.* 246–249.)

LAW OF VOLTA.

When any number of conductors, which conduct electricity without electrolytic dissociation, are joined together to form a closed chain, and all are at the same temperature, the total electromotive force, or sum of the contact-differences of potential at the surfaces of union of pairs of elements, is zero.

Thus, for the three elements a, b, c, which form a closed chain,

$$V_{ab} + V_{bc} + V_{ca} = 0,$$

where V_{ab} is the contact-difference of potential between the pair of elements a, b, and V_{bc}, V_{ca}, have similar meanings.

This result is known as Volta's Law.

(Consult Jeans, *Electricity and Magnetism,* p. 298. For a very full treatment, see Chwolson, *Traité de Physique,* Vol. IV, Part 10, pp. 198–200.)

KIRCHHOFF'S LAWS.

1. The algebraic sum of the currents which meet at any point is zero.
2. In any closed circuit the algebraic sum of the products of the current and resistance in each of the conductors in the circuit is equal to the electromotive force in the circuit.

(Thomson, *Elements of Electricity and Magnetism,* pp. 304, 305. For an illustration of the application of Kirchhoff's laws, see Northrup, *Methods of Measuring Electrical Resistance, art.* 301, pp. 44, 45.)

STEINMETZ'S EXTENSION OF KIRCHHOFF'S LAWS TO ALTERNATING CURRENTS.

(*a*) "The sum of all the e.m.fs acting in a closed circuit equals zero, if they are expressed by complex quantities, and if the resistance and reactance e.m.f.s are also considered as counter e.m.f.s."

(*b*) "The sum of all the currents flowing towards a distributing point is zero, if the currents are expressed as complex quantities."

(Steinmetz, *Alternating Current Phenomena, art.* 31, p. 40.)

RESOLVED ELECTROMOTIVE FORCE AND CURRENT.

(*a*) "The sum of the components, in any direction, of all e.m.f.s in a closed circuit, equals zero, if the resistance and reactance are considered as counter e.m.f.s."

(*b*) "The sum of the components, in any direction, of all the currents flowing to a distributing point equals zero."

(Joule's law and the energy-equation do not give a simple expression in complex quantities because power is a quantity of double the frequency of the current or e.m.f. wave.)

(Steinmetz, *Alternating Current Phenomena, art.* 31, p. 41.)

STEINMETZ'S EXTENSION OF OHM'S LAW TO ALTERNATING CURRENT.

$$\dot{E} = Z\dot{I},\ \dot{I} = \frac{\dot{E}}{Z},\ Z = \frac{\dot{E}}{\dot{I}},$$

where $\dot{E}$, $\dot{I}$ and Z are e.m.f., current and impedance, expressed in complex quantities.

(For a full explanation of these symbols and their relations, consult Steinmetz, *Alternating Current Phenomena,* Chap. V. See *art.* 30, p. 40.)

WORK DONE BY ELECTROMOTIVE FORCE.

The work done by an e.m.f. is measured by the product of the e.m.f. into the quantity of electricity which crosses a section of the conductor under the action of the e.m.f. This work is the same as the work done by an ordinary force and both are measured by the same standards or units.

(Maxwell, *Treatise on Electricity and Magnetism,* Vol. II, *art.* 570.)

POWER IN A CIRCUIT WHEN THE CURRENT IS ALTERNATING.

The *average power* in a circuit when the current is alternating (sine-waves assumed) is,

$$P = EI \cos \varphi,$$

where E is the effective value of the electromotive force, I the effective value of the current and φ the phase angle between the current and the impressed electromotive force.

The *instantaneous power* in a circuit is,

$$p = ei,$$

where e and i are the instantaneous values of the electromotive force and the current respectively.

(Steinmetz, *Alternating Current Phenomena, art.* 5, p. 6. Also consult Christie, *Electrical Engineering, art.* 80, pp. 113–118.)

LAW OF ACTION OF AN ELECTRIC CURRENT ON A MAGNET.

When current passes through a straight wire, a straight magnetized needle held above the wire tends to place itself at right angles to the wire and to the perpendicular let fall from the center of the magnetized needle to the wire. The direction of the current in the wire and the direction of the lines of magnetic force which encircle the wire are related as are the forward thrust and rotation of a right-handed screw.

(Consult *Ganot's Physics, art.* 835.)

ELECTRIC CIRCUIT AND A MAGNETIC SHELL COMPARED.

"A current flowing in any closed circuit produces the same magnetic field as a certain magnetic shell known as the 'equivalent magnetic shell.' This shell may be taken to be any shell having the circuit for its boundary, its strength being uniform and proportional to that of the current."

(See Jeans, *Electricity and Magnetism,* p. 415.)

AMPERE'S LAW FOR THE MAGNETIC FIELD DUE TO ANY CLOSED LINEAR CIRCUIT.

At any point P, not in the wire of a closed circuit carrying an electric current, the magnetic force due to the current can be derived from a potential Ω where Ω = a constant × the current × the solid angle subtended by the circuit at P. When electromagnetic measure is used the constant is unity, and $\Omega = i\,\omega$, i being the current flowing round the circuit and ω the solid angle subtended by the circuit at the point P.

(Thomson, *Elements of Electricity and Magnetism,* p. 325.)

FORCE ON A UNIT POLE EXTERIOR TO A LINEAR CONDUCTOR.

The force exerted by a linear conductor, carrying a current I on a unit magnetic pole *exterior* to the conductor, is perpendicular to the plane through the axis of the conductor and the pole.

When electromagnetic units are used this force is equal to twice the intensity of the current divided by the perpendicular distance r from the pole to the axis of the conductor.

$$\text{Thus, } T_0 = \frac{2I}{r}.$$

Here, T_0 is the force in dynes which would act on a unit magnetic pole, or it is the intensity of the magnetic field, at a distance r from the axis of the conductor.

(Consult "Some Newly Observed Manifestations of Forces in the Interior of an Electric Conductor," by E. F. Northrup, *Physical Review,* June, 1907, p. 478. Also Maxwell, *Treatise on Electricity and Magnetism, arts*. 477–479.)

MAGNETIC FORCE IN THE INTERIOR OF A CONDUCTOR OF CIRCULAR CROSS-SECTION.

The lines of magnetic force form circles about the axis of a linear conductor of circular cross-section. They vanish at the axis and are a maximum at the circumference of the conductor and increase uniformly with the distance from axis to circumference.

If electromagnetic units are used, the intensity of the magnetic field in the substance of the conductor is,

$$T_1 = \frac{2Ir}{R^2},$$

where I is the current in the conductor,

R the radius of the conductor and

r the distance from the axis to where the intensity of the magnetic force is T_1.

(Consult Maxwell, *Treatise on Electricity and Magnetism,* Vol. II, *art.* 683.)

A SMALL ELECTRIC CIRCUIT COMPARED WITH A MAGNET.

"The magnetic action of a small plane-circuit at distances which are great compared with the dimensions of the circuit is the same as that of a magnet whose axis is normal to the plane of the circuit, and whose magnetic moment is equal to the area of the circuit multiplied by the strength of the current."

(Maxwell, *Treatise on Electricity and Magnetism,* Vol. II, *art.* 482.)

MAGNETIC POLE AND EQUIPOTENTIAL SURFACES.

"The force acting on a magnetic pole placed at any point of an equipotential surface is perpendicular to this surface, and varies inversely as the distance between consecutive surfaces."

(See Maxwell, *Treatise on Electricity and Magnetism,* Vol. II, *art.* 487.)

LAW OF THE INTERNAL PRESSURE PRODUCED BY AN ELECTRIC CURRENT IN A CONDUCTOR.

In every conductor which carries an electric current, a pressure is produced within the substance of the conductor which results from the mutual attractions of all the current-carrying elements of the conductor.

When the conductor has a circular cross-section, the pressure is directed toward its axis. Its value per unit area is a maximum at the axis and decreases to zero at the circumference. When the radius of the conductor is R, and the current it carries is I, the pressure g per unit area at any radial distance r from the axis is,

$$g = \frac{I^2}{\pi R^4}(R^2 - r^2).$$

When I is in electromagnetic measure and R and r are in centimeters g is in dynes per cm^2.

(See "Some Newly Observed Manifestations of Forces

in the Interior of an Electric Conductor," by E. F. Northrup, *Physical Review*, June, 1907.)

LONGITUDINAL MOTION IN AN ELECTRICAL CONDUCTOR.

When by any geometrical disposition whatever of an electric circuit, in which the conducting material is a fluid capable of free motion, normally acting electrodynamic forces arise in any section of the circuit which vary in magnitude from one point to another over a length measured along the axis of the conductor, there also arise hydrodynamic forces which can impress motions on the fluid substantially parallel to the longitudinal axis of the conductor.

(This generalization is drawn from investigations by E. F. Northrup which are not yet published.)

WORK DONE IN MOVING A MAGNETIC POLE ROUND A CLOSED CURVE.

If there exists a magnetic field due to electric currents and a closed curve is drawn in this field, the work done in moving a magnetic pole of strength m round the closed curve is zero if the closed curve does *not* thread an electric circuit, and 4πm times the current in any circuit which the closed curve threads once.

Thus, $W = 4\pi Im$ is the work done when the closed curve threads once a circuit which carries a current I, and $W_1 = 4\pi Imn$ when the closed curve threads the circuit n times. The value of the line-integral $4\pi I$ is independent of the medium in which the closed curve is drawn.

(Maxwell, *Treatise on Electricity and Magnetism*, Vol. II, *arts.* 480, 498, 499.)

FUNDAMENTAL LINE-INTEGRALS.

The two propositions:

1. The line-integral of the magnetic force round any closed curve is equal to 4π times the total current flowing through the closed curve and
2. The line-integral of the electric force round any closed curve is equal to the time-rate of diminution of the total magnetic induction included by the closed curve,

are basic in many mathematical investigations of alternating currents.

(For application of these principles, see "The Skin Effects and Alternating Current Resistance," by E. F. Northrup and John R. Carson, *Jour. of the Franklin Institute,* Feb., 1914, p. 141 *et seq.*)

VECTORIAL ADDITION OF MAGNETIC AND ELECTRIC FORCES.

Magnetic and electric forces or intensities in a homogeneous medium must add vectorially, giving a resultant intensity.

By combining this principle with certain elementary laws of electrostatics and magnetism, a large number of theorems have been demonstrated.

(Consult Jeans, *Electricity and Magnetism,* pp. 26, 358.)

INTERACTION OF MAGNETS AND ELECTRIC CURRENTS.

The mechanical action of currents on magnets is equal and opposite to the action of magnets on currents.

It is shown by theory and experiments that a bar-magnet magnetized to have a North pole at each end and a South pole at its center, when freely suspended in a horizontal position in a vertical, conducting, liquid column of circular cross-section will rotate with continuous rotation when the liquid carries an electric current. Reversing the direction of the current reverses the direction of rotation.

(*Ganot's Physics, arts.* 888, 889. Also see article by E. F. Northrup, *Physical Review,* June, 1907, p. 480.)

ELECTRIC AND MAGNETIC ANALOGIES.

Electric system.

1. The line-integral of the electric force round any closed curve passing through the battery is E, while round any other closed curve it vanishes.
2. The lines of flow of electric current are closed curves which pass through the battery.
3. The density of the electric current is σ times the electric force, where σ is the conductivity of the medium carrying the current.

Magnetic system.

1. The line-integral of the magnetic force round any closed curve which threads a magnetizing circuit is $4\pi I$, while round any other closed curve it vanishes.
2. The lines of magnetic induction are closed curves which thread the magnetizing circuit.
3. The magnetic induction is μ times the magnetic force, where μ is the magnetic permeability.

(For amplification and applications of these analogies, see Thomson, *Elements of Electricity and Magnetism,* pp. 348–350.)

LAW OF MAGNETIC INDUCTION. (I)

"When the number of lines of magnetic induction which pass through the secondary circuit in the positive direction is altered, an electromotive force acts round the circuit, which is measured by the rate of decrease of the magnetic induction through the circuit"—provided the integrity of the original circuit is preserved.

(Maxwell, *Treatise on Electricity and Magnetism,* Vol. II, *art.* 531).

LAW OF MAGNETIC INDUCTION. (2)

Another, and to many, a preferable statement of this law is: Whenever a real or imaginary line in space is being cut at right angles to itself by tubes of magnetic induction an e.m.f. acts along this line which is proportional to the rate of cutting.

(Consult Maxwell, *Treatise on Electricity and Magnetism,* Vol. II, *art.* 541.)

LENZ'S LAW.

When a circuit is moved in a magnetic field in such a way that a change takes place in the number of tubes of magnetic induction passing through the circuit, a current is induced in the circuit and a mechanical force is set up such that this force tends to stop the motion which gave rise to the current.

(Thomson, *Elements of Electricity and Magnetism,* p. 440. Also Maxwell, *Treatise on Electricity and Magnetism,* Vol. II, *art.* 542.)

LINE OF MAGNETIC INDUCTION DEFINED.

Maxwell defines, negatively, a line of magnetic induction in four ways:

1. If a conductor be moved along it parallel to itself it will experience no electromotive force.
2. If a conductor carrying a current be free to move along a line of magnetic induction it will experience no tendency to do so.
3. If a linear conductor coincides in direction with a line of magnetic induction and be moved parallel to itself in any direction, it will experience no electromotive force in the direction of its length.
4. If a linear conductor carrying an electric current coincide in direction with a line of magnetic induction it will not experience any mechanical force.

(Maxwell, *Treatise on Electricity and Magnetism,* Vol. II, *art.* 597.)

ELECTROMOTIVE FORCE INDEPENDENT OF THE NATURE OF THE CONDUCTOR.

The intensity of an electromotive force which results from electromagnetic induction is entirely independent of the nature of the substance of the conductor in which the electromotive force acts, and also of the nature of the conductor which carries the inducing current.

(Maxwell, *Treatise on Electricity and Magnetism,* Vol. II, *art.* 534.)

MECHANICAL FORCE ACTS ON THE CONDUCTOR, NOT ON THE CURRENT.

The mechanical force which urges a conductor carrying a current across lines of magnetic force, acts, *not on the electric current* but on the *conductor* which carries it.

(Maxwell, *Treatise on Electricity and Magnetism,* Vol. II, *art.* 501.)

MECHANICAL FORCE ACTING ON A CONDUCTOR REPRESENTED BY A PARALLELOGRAM.

The mechanical force which acts upon unit length of a conductor carrying a current is numerically equal to the area of a parallelogram, two sides of which are drawn parallel to the conductor and proportional to the strength of the current at any point, the other two sides being drawn parallel and proportional to the magnetic induction at the same point. The mechanical force is normal to the plane of the parallelogram so drawn. If a right-handed screw be turned from the direction of the current to the direction of the induction the direction in which the mechanical force acts coincides with the direction of forward motion of the screw.

(Maxwell, *Treatise on Electricity and Magnetism,* Vol. II, *art.* 500.)

MAGNETIC ENERGY IS POTENTIAL ENERGY.

The energy of any strictly magnetic system may be considered as potential energy and if so considered this energy is always *diminished* when the parts of the system yield to the magnetic forces which act on them.

(Maxwell, *Treatise on Electricity and Magnetism,* Vol. II, *art.* 638.)

MUTUAL ACTION BETWEEN TWO CIRCUITS IS DEPENDENT UPON A SINGLE QUANTITY.

All phenomena of the mutual action of two circuits, whether the induction of currents or the mechanical force which acts between them, depend upon the value of the *coefficient of mutual induction* between the circuits, a coefficient which depends only upon the geometrical relations of the circuits.

(Maxwell, *Treatise on Electricity and Magnetism,* Vol. II, *art.* 584. Very exact formulæ for calculating coefficients of self induction and mutual induction are to be found in a series of papers by Dr. E. B. Rosa in volumes 2, 3, and 4 of the *Bulletin of the Bureau of Standards.*)

VECTOR RELATIONS OF VELOCITY, INDUCTION AND ELECTROMOTIVE FORCE.

"The magnitude of the electromotive force is represented by the area of the parallelogram, whose sides represent the velocity and the magnetic induction, and its direction is the normal to this parallelogram, drawn so that the velocity, the magnetic induction and the electromotive force are in right-handed cyclical order."

(Maxwell, *Treatise on Electricity and Magnetism,* Vol. II, *art.* 599. Also see *art.* 594, where is shown how a *volume* may be made to represent the increment in the electrokinetic momentum of a secondary circuit.)

INDUCTION COEFFICIENTS.

In every electric circuit there is a certain quantity L, called the *coefficient of self induction* of the circuit. The magnitude of this quantity is dependent only upon the geometrical dimensions of the circuit. If the currents are constant, changes in L always give rise to e.m.f.s, and if L is constant changes in the currents give rise to e.m.f.s. If two or more circuits are in the neighborhood of each other, there is for each of the circuits a self-induction coefficient and in addition other quantities M, M_1, M_2, etc., called coefficients of *mutual induction* which depend only upon the geometrical relations of the circuits. If there is a steady current in one of the circuits, and there are changes in L, L_1, L_2, etc., or in M, M_1, M_2, etc., e.m.f.s are always produced. Or if these quantities are constant, changes in the current produce e.m.f.s.

(Consult Maxwell, *Treatise on Electricity and Magnetism,* Vol. II, *arts.* 540, 578, 579.)

COIL TO GIVE MAXIMUM SELF INDUCTION.

When the weight or length of a wire is given, the form in which to wind this wire in a channel of square cross-section, in order to form a coil of maximum self induction, is obtained by making the mean radius r of the coil equal 1.85 times a side of the section of the channel.

The self induction in henrys is then,

$$L = 19.347rn^2\ 10^{-9},$$

where n is the number of turns of wire.

(Maxwell, *Treatise on Electricity and Magnetism,* Vol. II, *art.* 706. Also *Bulletin of the Bureau of Standards,* Vol. 2, p. 108.)

EXPRESSION FOR KINETIC ENERGY OF TWO CIRCUITS.

The kinetic energy of a system formed of two circuits when currents in the first circuit induce currents in the second circuit is given by the expression,

$$T = \frac{1}{2} I_1^2 L + I_1 I_2 M + \frac{1}{2} I_2^2 N,$$

where L is the coefficient of self induction of first circuit and N of the second circuit, and M is the coefficient of mutual induction between the two circuits, and I_1 and I_2 are currents in the first and second circuits respectively.

(Maxwell, *Treatise on Electricity and Magnetism,* Vol. II, *art.* 581.)

MAGNETIC ENERGY COMPARED WITH ELECTROKINETIC ENERGY.

It is always possible to make an arrangement of infinitely small electric circuits which shall correspond in all respects to any magnetic system, provided that in calculating the potential we avoid passing through any of these small electric circuits with a line of integration.

(Maxwell, *Treatise on Electricity and Magnetism,* Vol. II, *art.* 637.)

VECTOR-POTENTIAL.

The *vector-potential* is a quantity which represents in direction and magnitude, the time-integral of the *intensity* of the *electromotive* force which a particle placed at a point in a magnetic field would experience if the current to which the magnetic field is due were suddenly stopped. It is identical with the electrokinetic momentum *at the point.*

All lines of magnetic induction through a closed curve in being removed are shown by Maxwell to equal the line-integral of the resolved part of the vector-potential taken round the curve. This line-integral, physically interpreted, is the total electrokinetic momentum of the closed curve or circuit, and when the line-integral is extended round a primary circuit it is numerically equal to the product of the self induction of the circuit by the current in the circuit. It is the analogue of mass × velocity, or momentum in mechanics.

(Maxwell, *Treatise on Electricity and Magnetism*, Vol. II, *arts*. 405, 590, 592.)

ELECTROMOTIVE FORCE IMPRESSED ON A CIRCUIT.

If an e.m.f. is momentarily impressed on *any* electric circuit three cases may exist:

1. The movement of electricity is retarded by an ohmic resistance only.
2. It is retarded by ohmic resistance and opposed by magnetic inertia.
3. It is retarded by ohmic resistance and opposed by magnetic inertia and by a counter e.m.f. which varies as a function of the impressed e.m.f.

In the 3rd case when the damping by ohmic resistance is negligible the electricity always tends to oscillate upon the sudden removal of the impressed e.m.f. and the period is $T = 2\pi \sqrt{LC}$, where L is the self induction and C the capacity of the circuit.

(Consult Bedell and Crehore, *Alternating Currents*, Part 1.)

AN ELECTROMOTIVE FORCE ACTS ONLY ON ELECTRICITY.

An electromotive force has of itself no tendency to cause the mechanical motion of any body, but acts only to cause a movement of electricity.

(Maxwell, *Treatise on Electricity and Magnetism*, Vol. II, *art.* 535.)

DECAY OF INTERNAL CHARGES IN DIELECTRICS.

If in the interior of a mass of homogeneous poorly conducting material, there exists at any point an electric charge it will tend to die away. Neither its formation nor the rate at which it dies away is influenced by the application of external e.m.f.s which do not lead to disruptive discharges.

(Maxwell, *Treatise on Electricity and Magnetism,* Vol. I, *art.* 325.)

GENERAL PRINCIPLE OF MECHANICAL ACTION OF CURRENTS.

All *mechanical* actions of electric currents depend upon the strength of the currents and not upon their rate of variation and all mechanical actions of currents remain the same when *all* the currents are reversed in direction.

(Maxwell, *Treatise on Electricity and Magnetism,* Vol. II, *art.* 574.)

MUTUAL RELATIONS OF CURRENTS.

1. If a current traverses a wire and returns by a tube surrounding the wire there is no magnetic field external to the tube.
2. The external action of a crooked wire (bent like a row of saw-teeth) upon a neighboring wire is the same as that of a straight wire.
3. If a wire carries a current no external magnetic force can so act upon the wire as to tend to make it move in the direction of its length.
4. The force acting between two elements of two electric circuits is inversely proportional to the square of the distance between them.

(Maxwell, *Treatise on Electricity and Magnetism,* Vol. II, *arts.* 505–508.)

INTERACTION OF ELECTRIC CONDUCTORS.

1. Two conductors which are parallel and carry currents in the same direction attract one another, and when

they carry currents in opposite directions they repel one another.

2. Two rectilinear current-carrying conductors, when their directions are such that they form an angle with each other, attract one another if the currents in both conductors approach or recede from the apex of the angle, and they repel one another if one current approaches and the other recedes from the apex of the angle. (*Ganot's Physics, arts.* 880–882.)

LAW OF RESOLUTION OF CURRENTS.

The law of the resolution of electric currents is the same as that of velocities, forces and all other vectors.

(Maxwell, *Treatise on Electricity and Magnetism,* Vol. I, *art.* 286.)

MAGNETOMOTIVE FORCE, AND MAGNETIC INDUCTION IN A SOLENOID.

In a solenoid of infinite length having n turns per unit length, the magnetic force is uniform over the cross-section of the solenoid and its value is $H = 4\pi ni$, where i is the current in electromagnetic measure. Or, if I is the current in amperes the magnetomotive force is

$$H = 1.2566nI \text{ gilberts.}$$

The induction per cm^2 in the solenoid is,

$B = 1.2566nI\mu$ gausses, where μ is the permeability of the medium within the solenoid.

(Consult Christie, *Electrical Engineering,* p. 58 *et seq.*)

A FUNDAMENTAL ENGINEERING EQUATION.

The following equation is of fundamental importance in electrical engineering:

$$E_{eff} = \sqrt{2}\,\pi n\,\varphi N 10^{-8} = 4.44 n\,\varphi N 10^{-8} \text{ volts.}$$

Here, E_{eff} = effective e.m.f., n = total number of turns in the circuit, φ = the total maximum flux through the circuit and N = the frequency or number of complete cycles per second of the magnetizing current, sine waves being assumed.

(Steinmetz, *Alternating Current Phenomena,* Chap. III. See p. 17.)

THERMOELECTRIC CURRENTS.

If an electric circuit is made up of two unlike metals and one junction of the two metals is maintained at a higher temperature than the other junction, an electric current will flow in the circuit, and as the result of this current heat will be transferred from the hotter toward the colder junction.

(Thomson, *Elements of Electricity and Magnetism,* p. 501. Also Chwolson, *Traité de Physique,* Vol. IV, Part 10, p. 737.)

THERMOELECTRIC LAW, THERMOELECTRIC POWER.

If e_1 is the e.m.f. of a bimetallic circuit when the cold junction is at temperature t_0 and the hot junction at t_1 and if e_2 is the e.m.f. when the cold junction is at t_1 and the hot junction at t_2, then, when the cold junction is at t_0 and the hot junction is at t_2 the e.m.f. is $e_1 + e_2$. Or, the e.m.f. E round a circuit whose junctions are at temperatures t_1 and t_2 is $E = \int_{t_1}^{t_2} Q\,dt$, where Qdt is the e.m.f. round the circuit when the temperature of the cold junction is $t - \frac{1}{2}\,dt$ and the temperature of the hot junction is $t + \frac{1}{2}\,dt$. The quantity Q is known as the *thermoelectric power* of the circuit at temperature t.

(Thomson, *Elements of Electricity and Magnetism*, p. 506. Also consult Chwolson, *Traité de Physique*, Vol. IV, Part 10, p. 744. For values of thermoelectric power see *Smithsonian Physical Tables*, pp. 268–270.)

THERMOELECTRIC LAW FOR DIFFERENT PAIRS OF METALS.

If E_{ac} is the e.m.f. round a circuit formed of the pair of metals A, C, and E_{bc} the e.m.f. round a circuit formed of the pair of metals B, C, then $E_{ac} - E_{bc}$ is the e.m.f. round a circuit formed of the metals A and B; all these circuits being supposed to work between the same limits of temperature.

(Thomson, *Elements of Electricity and Magnetism*, p. 506.)

PELTIER EFFECT.

When current flows across the junction of two unlike metals it gives rise to an absorption or liberation of heat. If the current flows in the same direction as the current at the hot junction in a thermoelectric circuit of the two metals heat is absorbed; if it flows in the same direction as the current at the cold junction of the thermoelectric circuit heat is liberated. This phenomenon is known as the Peltier Effect.

(Thomson, *Elements of Electricity and Magnetism*, pp. 501, 502. Also *Ganot's Physics, art.* 878. Consult also *Smithsonian Physical Tables*, p. 271.)

MEASURE OF PELTIER EFFECT.

The Peltier Effect equals (the thermoelectric power) $\times$ (the absolute temperature). Or, $P = QT$, where $T =$ absolute temperature.

(Thomson, *Elements of Electricity and Magnetism*, p. 510. Also Maxwell, *Treatise on Electricity and Magnetism*, Vol. 1, *art.* 249.)

The Peltier Effect, the closely associated Thomson Effect and the Hall Effect are thought by many to be explained with fair satisfaction on the modern electron-theory. For advanced views on this matter consult Campbell, *Modern Electrical Theory*, pp. 71–74.)

ACTION OF CURRENT FLOWING ALONG AN UNEQUALLY HEATED CONDUCTOR, CALLED THE "THOMSON EFFECT."

When an electric current flows along an unequally heated metallic conductor it tends, in the case of copper, to diminish the inequality of temperature and in the case of iron to increase this inequality.

(Thomson, *Elements of Electricity and Magnetism*, p. 505. Also, *Maxwell, Treatise on Electricity and Magnetism*, Vol. I, *art.* 253. Also, for full account, see Chwolson, *Traité de Physique*, Vol. IV, Part 10, pp. 752–756.)

RICHARDSON'S LAW OF ELECTRONIC EMISSION.

"The number of electrons emitted at different temperatures T is governed by the formula,

$$N = A\, T^{\lambda} e^{-b/T}.$$

A, λ and b are constants." A varies greatly with the substance, λ is not far different than unity and b in equivalent volts is always comparable with five.

(Richardson, *The Electron Theory of Matter*, p. 441. Also O. W. Richardson, *Phil. Trans.* (A), Vol. CCI, p. 543, 1903.)

LAW OF MAGNUS.

If a circuit is formed of a single metal no current will be formed in it however the section of the conductor and the temperature may vary in different parts of the circuit.

(Maxwell, *Treatise on Electricity and Magnetism,* Vol. I, *art.* 251.)

THERMOELECTRIC INVERSION.

Call E the e.m.f. acting round a thermoelectric circuit of two metals. Then, when the difference of temperature of the two junctions is twice the difference of temperature for which E is a maximum, E becomes zero and a further increase in the temperature-difference, the lower temperature of the one junction being held constant, causes E to change direction. This is known as *thermoelectric inversion.*

(See Chwolson, *Traité de Physique,* Vol. IV, Part 10, p. 739. Thermoelectric phenomena are very adequately and fully treated in Chap. VI, pp. 728–761.)

RATIO OF THE ELECTROMAGNETIC TO THE ELECTROSTATIC UNIT OF A QUANTITY OF ELECTRICITY.

The ratio of the *electromagnetic* to the *electrostatic unit* of a quantity of electricity is a velocity (whatever units of length and time are chosen) and this velocity is experimentally shown to be the velocity of light. The ratio is called v and, very nearly,

$$v = 3 \times 10^{10} \text{ cm/sec.}$$

(Consult Maxwell, *Treatise on Electricity and Magnetism,* Vol. II, Chap. XIX. Also Thomson, *Elements of Electricity and Magnetism,* p. 470 *et seq.*)

RELATION OF SPECIFIC INDUCTIVE CAPACITY AND INDEX OF REFRACTION.

According to Maxwell's electromagnetic theory of light, the dielectric capacity (specific inductive capacity) of a transparent medium is equal to the square of its index of refraction for electromagnetic radiation.

Thus, $K = \mu^2$.

(Consult Maxwell, *Treatise on Electricity and Magnetism,* Vol. II, *arts.* 788, 789.)

ROTATION OF THE PLANE OF POLARIZATION OF LIGHT.

In 1845 Faraday discovered that if a substance which ordinarily will not rotate plane-polarized light be placed in a strong magnetic field it acquires this property. In looking from North to South along a line of magnetic force the rotation is clockwise. On reversing the direction of magnetization, the direction of rotation is reversed.

The angular rotation of the plane of polarization of a plane-polarized ray of light, which is parallel to a magnetic field, is numerically equal to the amount by which the magnetic potential increases in passing from the point where the ray enters the medium, (in which the rotation takes place) to the point where the ray leaves it, multiplied by a coefficient. This coefficient is generally positive for diamagnetic media.

(See Preston, *The Theory of Light,* p. 431. Also Maxwell, *Treatise on Electricity and Magnetism,* Vol. II, *art.* 808. For formulæ and numerical values of Verdet's constant see *Smithsonian Physical Tables,* pp. 326–330.)

PRESSURE OF RADIANT ENERGY.

In a medium, in which radiant energy is propagated as light, heat or electromagnetic waves, there is a *pressure* in the direction normal to the wave front. This pressure is numerically equal to the energy in unit volume of the medium.

If K is the specific inductive capacity and μ the permeability of the medium, P the maximum electromotive force and β the maximum magnetic force (the direction of which is at right angles to both the direction of P and the direction of the propagation of the wave) then,

$\frac{K}{8\pi}P^2 = \frac{\mu}{8\pi}\beta^2 =$ mean energy in unit volume of the medium.

(See Maxwell, *Treatise on Electricity and Magnetism,* Vol. II, *arts.* 792, 793. For description of Nichols and Hull's experimental proof of the above, see Wood, *Physical Optics,* p. 466.)

HALL EFFECT.

When a thin rectangular sheet of metal carrying an electric current flowing in the direction of its length is subjected to a powerful magnetic field normal to the sheet, the current stream-lines are deflected toward one edge of the sheet. This is called the Hall Effect.

(*Ganot's Physics, art.* 900. Also Campbell, *Modern Electrical Theory,* pp. 76–78.)

ELECTRO-OPTICAL EFFECT IN DIELECTRICS.

When certain dielectrics are subjected to electric strain they become doubly refracting. Kerr states the law as follows:

"The strength of the electro-optical action of a given dielectric, that is the difference in the path of the ordinary and extraordinary rays, for unit thickness of the dielectric, varies directly as the square of the resultant electric force."

(*Ganot's Physics, art.* 997.)

LAW OF SARASIN AND DE LA RIVE, OF "MULTIPLE RESONANCE."

The distance between two nodes on a resonator changes in changing the resonator but not in changing the oscillator. This distance (the internode) is the half wave-length of the free oscillations of the resonator only. (These results are connected with the rapid damping of the waves which usually occurs on the oscillator.)

(Vreeland, *Maxwell's Theory of Wireless Telegraphy*, p. 62.)

LAW OF DISTRIBUTION OF ELECTRO-MAGNETIC RADIATION. (1)

At any point at a distance from the center of disturbance there is an electric and magnetic disturbance at right angles to the line drawn from the center of disturbance. The electric force is on a tangent to a great circle of a sphere with the oscillator at its center and which has its poles at the intercepts of the axis of the oscillator produced: and the magnetic disturbances lie on tangents to circles parallel to a plane perpendicular to the oscillator.

(Vreeland, *Maxwell's Theory of Wireless Telegraphy*, pp. 76, 77.)

LAW OF DISTRIBUTION OF ELECTRO-MAGNETIC RADIATION. (2)

The two vibrations (electric and magnetic) are transverse, as in light, and perpendicular to the direction of propagation of the wave-front. The amplitude of these vibrations varies inversely as the distance and their intensity varies inversely as the square of the distance from the oscillator. The vibrations maintain a constant direction as do those of polarized light.

(Vreeland, *Maxwell's Theory of Wireless Telegraphy*, pp. 76, 77.)

RELATION OF MAGNETIC FORCE AND A MOVING FARADAY TUBE.

A Faraday tube in motion perpendicular to its direction always gives rise to a magnetic force, the direction of motion, the Faraday tube and the magnetic force being mutually at right angles, when the medium is isotropic.

(Thomson, *Elements of Electricity and Magnetism*, pp. 479, 490–492.)

POYNTING'S LAW.

When a conductor carrying a current is in an electrostatic field the transfer of energy takes place through the dielectric along paths which are the intersections of the equipotential surfaces of the electrostatic field with the equipotential surfaces of the electromagnetic field due to the current.

(Gray, *A Treatise on Magnetism and Electricity*, p. 421.)

GENERALITY OF LAW OF INVERSE-SQUARES.

It may be stated generally that the intensity of an effect, which emanates from a center and is transmitted equally in all directions, is, in an isotropic medium, inversely proportional to the square of the distance from the source.

(New Century Dictionary under word LAW.)

VI
LIGHT

LIGHT.

A GENERALITY IN RADIATION.

Every fact of experience and every consideration of theory go to prove that radiant energy of all wave-lengths, whether called electric, heat-, or light-radiation, travels in space free from ordinary matter with the same velocity and obeys the same laws of propagation. This is a velocity which would be determined on an elastic solid theory by the equation

$$v = \sqrt{\frac{e}{d}},$$

where e is the elasticity and d the density of the ether of space.

(Consult Chwolson, *Traité de Physique*, Vol. II, Part 8, pp. 2 and 94 *et seq.* Also Preston, *The Theory of Light*, pp. 28–30.)

LIGHT DEFINED: VELOCITY OF.

Light consists of radiant energy, propagated in free space with the velocity common to all radiant energy but having wave-lengths such that it affects the human eye.

Calling λ the wave-length and N the frequency of the transverse vibrations the velocity is very approximately $v = \lambda N = 3 \times 10^{10}$ cms. per second = 300,000 kilometers per sec. The wave-length of visible radiant energy lies within the limits $\lambda = 0.76\mu$ and $\lambda = 0.4\mu$, and the frequency lies in the limits $N = 4 \times 10^{14}$ and $N = 7.5 \times 10^{14}$. (μ = 0.001 millimeter.)

(Consult Chwolson, *Traité de Physique*, Vol. II, Part 8, p. 29. For methods of determining the velocity of light, see Preston, *The Theory of Light*, Chap. XIX, p. 489 *et seq.*)

RECTILINEAR PROPAGATION OF LIGHT.

Light travels in straight lines through a homogeneous medium if the rays are not compelled to pass through any very small openings. When it passes through media of different kinds, it does not, in general, travel in the same straight line through them all.

(Crew, *General Physics*, p. 428. Also *Ganot's Physics, art.* 513.)

INTENSITY OF RADIATION.

The quantity of energy which traverses in the unit of time unit surface normal to the ray is here chosen to define the *intensity of radiation.* This intensity is inversely proportional to the square of the distance to a point-source of the radiation. The intensity of radiation received on an oblique surface is proportional to the cosine of the angle which the ray makes with the normal to the illuminated surface.

(Chwolson, *Traité de Physique*, Vol. II, Part 8, pp. 24, 25. Also *Ganot's Physics, art.* 519.)

LAWS OF REFLECTION.

1. The angle of reflection is equal to the angle of incidence.

2. The incident and the reflected ray are both in the same plane which is perpendicular to the reflecting surface.

(*Ganot's Physics, art.* 522. Also Preston, *The Theory of Light*, p. 74. For extensive treatment consult Chwolson, *Traité de Physique*, Vol. I, Part 2, p. 168 and Vol. II, Part 8, p. 112 *et seq.*)

REFLECTION FROM A PLANE-MIRROR.

"The image of a point in a plane-mirror lies on the perpendicular let fall from the point to the mirror, and lies as far behind the mirror as the point lies in front of the mirror."

(Crew, *General Physics,* p. 441. Also *Ganot's Physics, art.* 524.)

REAL AND VIRTUAL IMAGES.

Real images are those formed by the reflected rays themselves, and virtual images are those formed by their prolongations. Real images can be received on a screen, virtual images cannot. Plane-mirrors give rise to virtual images only.

(*Ganot's physics, art.* 525.)

LIGHT REFLECTED FROM A ROTATING MIRROR.

If a ray of light falls on a plane mirror and the mirror is rotated through any angle about an axis which lies in the plane of the mirror, the reflected ray is rotated through twice the angle.

(Kimball, *College Physics,* p. 563. Also *Ganot's Physics, art.* 530.)

LAW OF FERMAT, OR PRINCIPLE OF LEAST TIME.

When light passes from any point P to another point P′ by reflection at a point p on a surface, the path P p P′ is that which will be traversed by the ray in the least time in passing from P to P′ by reflection at the surface. A similar law applies for the paths of refracted rays, so that when light travels from one point to another the ray pursues that path which requires the least time.

(Preston, *The Theory of Light,* p. 95. Also Chwolson, *Traité de Physique,* Vol. II, Part 8, pp. 114, 115.)

REFLECTIONS FROM PORTIONS OF THE SURFACE OF A SPHERE.

The formulas for calculating the location of images formed by spherical mirrors are simple only when the spherical surface is but a small portion of the surface of a sphere. In this case the distance s from a point-source to a point on the surface of the mirror, the distance s′ from the image to the same point on the surface, and the radius r of the spherical mirror bear the relations,

$$\frac{s}{s'} = \frac{s-r}{r-s'}, \text{ or } \frac{1}{s} + \frac{1}{s'} = \frac{2}{r}.$$

These general relations hold for both concave and convex surfaces which are small portions of a sphere, if, calling the center of the mirror-surface the origin, distances to the right are reckoned positive and to the left negative.

(Chwolson, *Traité de Physique*, Vol. II, Part 8, pp. 115–122. Also Crew, *General Physics*, pp. 445, 446.)

RELATIVE SIZE OF OBJECT AND IMAGE.

For a spherical mirror,

$$\frac{\text{the linear dimensions of the image}}{\text{the linear dimensions of the object}} = \frac{\text{the distance from a point on the mirror to the image}}{\text{the distance from the same point on the mirror to the object}}.$$

This relation holds whether the image be real or virtual and for convex as well as concave mirrors.

(*Ganot's Physics, art.* 540.)

THE CAUSTIC.

When the aperture of a spherical mirror much exceeds 10 degrees the rays from a point-source reflected by the mirror suffer spherical aberration by reflection. Every reflected ray cuts the one adjacent to it and their points of intersection form in space a curved surface which is called the *caustic by reflection.*

(*Ganot's Physics, art.* 542.)

REFLECTION FROM A PARABOLIC MIRROR.

In reflection from a parabolic mirror all rays parallel to its axis, after reflection, meet at the focus of the mirror; and conversely, when a point-source of light is placed at the focus, the rays incident on the mirror are reflected exactly parallel to the axis, and their intensity tends to remain constant at all distances.

(*Ganot's Physics, art.* 544.)

REFLECTION FROM MAT SURFACES: LAMBERT'S LAW.

Surfaces from which the diffusion of light is uniform are said to be *mat*.

The reflection, from such surfaces, of light incident under a given angle follows the law of cosines, often called "Lambert's law." According to this law the intensity of the light diffusely reflected from a mat surface is proportional to the cosine of the angle between the direction of the diffused rays under consideration and the normal to the surface. However, if the angle of incidence be varied, the light reflected at a constant angle to the normal is not proportional to the angle of incidence.

(See *New Century Dictionary,* under title COSINE LAW, and the sub-head, *Lambert's Law of Cosines.*)

SELECTIVE REFLECTION.

Selective reflection is reflection in which the incident and reflected rays differ in composition. Many otherwise transparent bodies have absorption-bands in their spectra, and waves of light having a wave-length to which the body is opaque are more completely reflected than the others.

(*New Century Dictionary,* under word REFLECTION. Consult Wood, *Physical Optics,* pp. 352, 353.)

REFLECTION FROM AN ELEMENT OF A NON-SPHERICAL SURFACE.

A bundle of rays, incident normally on an infinitely small area of a non-spherical surface, give in reflection two infinitely small right-angled rectilinear focal lines, parallel to the elements of lines of curvature of this surface. A bundle of such rays is called *astigmatic.*

(Chwolson, *Traitê de Physique,* Vol. II, Part 8, pp. 125, 126. See also Preston, *The Theory of Light,* p. 109.)

VELOCITY OF LIGHT IN ORDINARY MATTER.

Light travels more slowly in any kind of ordinary matter than in vacuum. Thus for red rays,

$$\frac{\text{velocity in air}}{\text{velocity in water}} = 1.329 \text{ (about)}$$

$$\text{and } \frac{\text{velocity in air}}{\text{velocity in carbon bisulphide}} = 1.612 \text{ (about)}$$

(*Ganot's Physics, art.* 561. Also Ames, *Theory of Physics,* pp. 424–426.)

LAWS OF REFRACTION: SNELL'S LAW.

"The incident and refracted rays are in the same plane with the normal to the surface; they lie on opposite sides of it, and the sines of their inclinations to it bear a constant ratio to one another."

Denoting the angles of incidence and refraction by i and r respectively, the relation between them is given by the formula,

$$\frac{\sin i}{\sin r} = \mu.$$

The constant ratio μ is called the *index of refraction.*

(Preston, *The Theory of Light,* p. 88.)

RELATION OF VELOCITIES AND INDEX OF REFRACTION.

For any particular medium the index of absolute refraction μ varies inversely as the velocity of light in that medium.

Denoting by i and r the angles of incidence and refraction respectively, and by v and v′ the velocities in ether and the medium respectively,

$$\frac{\sin i}{\sin r} = \frac{v}{v'} = \mu.$$

(Preston, *The Theory of Light,* pp. 89, 90.)

RELATIONS OF VELOCITIES IN MEDIA OTHER THAN ETHER.

If the velocities in two media are v_1 and v_2 while the velocity in ether is v, then $\mu_1 = \frac{v}{v_1}$ and $\mu_2 = \frac{v}{v_2}$, or,

$$\frac{v_1}{v_2} = \frac{\mu_2}{\mu_1}.$$

Also $\mu_1 \sin i = \mu_2 \sin r$.

For any number of media the continued product of the relative refractive-indices of n substances is equal to the ratio of the absolute refractive-index of the nth substance to that of the first. Or in a formula,

$$\mu_{12} \cdot \mu_{22} \cdot \cdots \cdot \mu_{n-1} \cdot \mu_n = \frac{\mu_n}{\mu_1}$$

(Preston, *The Theory of Light,* pp. 90, 91. Also Ames, *Theory of Physics,* p. 425.)

ATMOSPHERIC REFRACTION.

The refractive index of the earth's atmosphere decreases as we ascend and for this reason all light which reaches us from stars not in the zenith travels in curved paths. The effect of this refraction is to apparently raise the stars toward the zenith. The mirage results from the double view of an object given by rays reaching the eye by two paths, one nearly direct and one concave upward, giving the effect of reflection.

(*Ganot's Physics, art.* 551. Also Wood, *Physical Optics,* pp. 69, 70.)

TOTAL REFLECTION AND CRITICAL ANGLE.

Let μ_1 and μ_2 be the indices of refraction of two media, μ_1 being greater than μ_2. Then, if a ray of light in passing through the medium of greater refraction comes to the boundary of the two media, it can only pass out into the medium of less refraction when the sine of the angle of incidence (angle between ray and normal to surface of separation of the two media) is less than $\frac{\mu_2}{\mu_1}$.

For all values of the angle of incidence greater than this the ray is totally reflected from the plane of separation of the two media.

The maximum value of the angle of incidence which does not give total reflection is known as the "critical" angle.

(Ames, *Theory of Physics*, p. 427. Also *Ganot's Physics, art.* 550.)

REFRACTION AT A SINGLE SPHERICAL SURFACE.

When light passes from a point-source in a medium of refractive index μ_1 into a medium of refractive index μ_2 bounded by a spherical surface, then,

$$\mu_1\left(\frac{1}{r} - \frac{1}{s}\right) = \mu_2\left(\frac{1}{r} - \frac{1}{s'}\right).$$

where r = radius of curvature of refracting surface — r being taken positive when convex side of surface is toward incident ray — s = distance from apex of surface to point-source — s being taken negative when on same side of surface as point-source — and s′ = distance from apex to image — where s′ is taken negative or positive according as it is on the same or opposite side of the surface as the point-source.

(Crew, *General Physics*, p. 458.)

POSITION OF IMAGE FORMED BY REFRACTION AT A SPHERICAL SURFACE

The position of the image when rays pass from air into a substance bounded by a spherical surface is given by the formula,

$$\frac{\mu}{s'} - \frac{1}{s} = \frac{\mu - 1}{r},$$

where μ = index of refraction of substance used, s' = distance from apex of surface to image, s = distance from apex of surface to source and r = radius of curvature of spherical surface.

(This formula comes from the one just preceding by placing $\mu_1 = 1$ and $\mu_2 = \mu$.)

(Crew, *General Physics*, p. 460. Also *Ganot's Physics*, *art*. 552.)

GENERAL EQUATION FOR THIN LENSES.

$\frac{1}{p'} - \frac{1}{p} = (n-1)\left(\frac{1}{r} - \frac{1}{s}\right)$ is given in *Ganot's Physics*, *art*. 563, as a general formula applicable to all cases both of convex and concave lenses.

The point where the refracted ray cuts the axis when the incident ray is parallel to the axis is called the *principal focus*. Its distance from the lens is the focal length of the lens. The power of a lens equals $\frac{1}{\text{its focal length}}$. The unit of power is the " dioptric " which equals the power of a lens of focal length of one meter.

(See above reference for interpretation of this formula, or consult Crew, *General Physics*, p. 466.)

CHROMATIC ABERRATION.

The index of refraction for a given substance is not the same for all wave-lengths, or colors of light. Thus,

μ for,	Red light, C.	Yellow light, D.	Green light, F.
Flint glass	1.630	1.635	1.848
Crown glass	1.527	1.530	1.536

This failure of a lens to bring all colors to the same focus is known as "Chromatic Aberration."

(Crew, *General Physics*, p. 470. Also *Ganot's Physics*, *art.* 594.)

DEVIATION OF A RAY IN PASSING THROUGH A PRISM.

The angle of deviation δ of a ray passing through a prism is the angle between the incident and emergent rays.

$\delta = (i_1 - r_1) + (i_2 - r_2)$.

Here i_1 = angle of incidence (angle between normal to first face of prism and incident ray),

i_2 = angle of emergence (angle between normal to second face of prism and emergent ray),

r_1 = angle of refraction at first face and

r_2 = angle of refraction at second face. Also angle of prism is, $\alpha = r_1 + r_2$.

The angle of deviation δ is a minimum when the incident ray makes the same angle with the first face of the prism as the emergent ray does with the second face of the prism.

(Crew, *General Physics*, p. 471.)

VELOCITY OF LIGHT IN MEDIA DEPENDS UPON WAVE-LENGTH.

The velocity of light is less the more highly refracting the medium, and as light of short wave-lengths is more refracted than light of long wave-lengths, the velocity of the former is less than that of the latter in all media of greater refractive index than unity.

(Consult Preston, *The Theory of Light*, Chap. XIX, pp. 513, 516. Also Wood, *Physical Optics*, pp. 15, 16.)

DISPERSIVE POWER.

The transformation of a beam of parallel rays of white light into a divergent pencil of light of different colors is knows as *dispersion*.

The *dispersive power* of a prism is the ratio of the angle of separation produced of two selected rays to the mean deviation (angle between incident and emergent rays) of the two rays.

$$P = \frac{d_h - d_a}{d} = \frac{n_h - n_a}{n - 1}.$$

Here n_h is the refractive index of the material of a prism or lens for violet rays, n_a the refractive index for red rays, n the refractive index for mean rays, and d_h,d_a, and d are the corresponding deviations.

(*Ganot's Physics, art.* 577.)

VARIATION OF REFRACTIVE INDEX WITH DENSITY (GLADSTONE AND DALE'S LAW).

When a substance is compressed or its temperature varied the density changes. This is accompanied by a corresponding variation in the refractive index such that,

$$\frac{\text{refractive index} - 1}{\text{density}} = \text{a constant}$$

(Preston, *The Theory of Light*, p. 131 *et seq.*)

BRIOT'S FORMULA.

Briot has determined the form of the function which expresses the variation of the index of refraction μ with wave-length λ. His formula is,

$$\frac{1}{\mu^2} = K\lambda^2 + A + \frac{B}{\lambda^2} + \frac{C}{\lambda^4} + \ldots\ldots,$$

where A, B, C, etc., are constants depending on the nature of the medium and diminishing rapidly as we proceed to higher terms, and K is another constant.

(Preston, *The Theory of Light*, pp. 487, 488.)

DOUBLE REFRACTION: OPTIC AXES.

In all transparent crystals, of which the fundamental form is not a cube, a black dot seen through the crystal appears double for most positions of the crystal. There are, however, in some crystals one and in others two directions, along which the dot being viewed appears single. This direction or directions constitute the optical axis or axes of the crystal.

This phenomenon, called double refraction, is very marked in the uniaxial crystal, Iceland spar (calcium carbonate.)

(Preston, *The Theory of Light*, p. 299. Also *Ganot's Physics, art.* 654.)

DOUBLE REFRACTION IN AN UNIAXIAL CRYSTAL: GENERAL STATEMENT.

Whatever be the plane of incidence, the ordinary ray always obeys the two general laws of single refraction.

In every section perpendicular to the optic axis, the extraordinary ray follows the laws of single refraction.

In every principal section the extraordinary ray follows the second law of refraction only, but the ratio of the sines of the angles of incidence and refraction is not constant.

(*Ganot's Physics, art.* 656.)

DOUBLE REFRACTION IN BIAXIAL CRYSTALS: GENERAL STATEMENT.

When a ray of light enters a biaxial crystal and traverses it in any direction not coinciding with an optic axis, it bifurcates and generally both rays are extraordinary rays. However, in a section of the crystal at right angles to the medial line one ray follows the laws of ordinary refraction, and in a section at right angles to the supplementary line the other ray follows the laws of ordinary refraction.

(*Ganot's Physics, art.* 658.)

CONICAL REFRACTION.

This is of two kinds, *internal* conical refraction and *external* conical refraction. For the former, an unpolarized ray entering a crystal along an axis of *single wave-velocity* diverges within the crystal as a cone and emerges as a hollow cylinder of rays. These can be received on a screen as a ring of light of constant diameter. In the latter case, if a conical pencil of rays enters a crystal with the apex of the cone on the face of the crystal and with its axis chosen parallel to the axis of single wave-velocity the emergent pencil of rays (superfluous rays being screened off) forms a hollow cone. This gives on a screen a ring of light which increases in diameter with the distance of the screen from the crystal.

(Wood, *Physical Optics,* pp. 257–259.)

POLARIZATION EFFECTED BY A CRYSTAL.

In traversing an uniaxial crystal the ordinary ray is polarized in the plane of the principal section, which includes this ray and the optic axis. The extraordinary ray, is on the contrary, polarized in a plane perpendicular to the plane of the principal section passing through the ray and the optic axis. The polarization is complete for both ordinary and extraordinary rays.

(Chwolson, *Traité de Physique,* Vol. II, Part 8, pp. 760, 761.)

POLARIZATION BY REFLECTION.

When a ray of ordinary light falls at an angle upon a reflecting surface of a transparent substance the reflected ray is more or less polarized. For a particular angle of incidence the polarization of the reflected ray is most complete and this angle is called the polarizing angle for the substance.

(*Ganot's Physics, art.* 667, or Wood, *Physical Optics,* Chap. IX. See p. 231.)

ANGLE OF POLARIZATION: BREWSTER'S LAW.

In polarization by reflection, the *polarizing angle* of a substance is that angle of incidence for which the reflected polarized-ray is at right angles to the refracted ray.

Brewster's law states that: The tangent of the polarizing angle for a substance is equal to the index of refraction of that substance. This law is expressed by the formula,

$$\mu = \tan i = \frac{\sin i}{\cos i}.$$

$$\text{As } \frac{\sin i}{\sin r} = \mu \text{ (Snell's law),}$$

where i is the angle of incidence of the reflected ray and r the angle which the refracted ray makes with the normal to the surface of separation of the two media, it follows that $i + r = 90°$; namely, the reflected and refracted rays form a right angle.

(*Ganot's Physics, art.* 668. Also Wood, *Physical Optics,* p. 231. Also Preston, *The Theory of Light,* pp. 303, 304.)

LAW OF MALUS.

When a pencil of light, polarized by reflection at one plane-surface, is allowed to fall upon a second plane-surface at the polarizing angle, the intensity of the twice-reflected beam varies as the square of the cosine of the angle between the two planes of reflection.

(Preston, *The Theory of Light*, p. 305. Also Wood, *Physical Optics*, p. 235 *et seq.*)

POLARIZATION OF REFRACTED LIGHT.

The relation between the polarized light in the refracted pencil and that in the reflected beam was discovered by Arago and is stated thus: "When an unpolarized ray is partly reflected at, and partly transmitted through, a transparent surface, the reflected and transmitted portions contain equal quantities of polarized light, and the planes of polarization are at right angles to each other."

(Preston, *The Theory of Light*, p. 304.)

LAWS OF INTERFERENCE OF POLARIZED LIGHT.

"1. Two rays of light polarized at right angles do not interfere destructively under the same circumstances as two rays of ordinary light.

2. Two rays polarized in the same plane interfere like two rays of ordinary light.

3. Two rays polarized at right angles may be brought to the same plane of polarization without thereby acquiring the quality of being able to interfere with each other.

4. Two rays polarized at right angles, and afterwards brought to the same plane of polarization, interfere like ordinary light if they originally belonged to the same beam of polarized light."

(Preston, *The Theory of Light*, p. 308.)

THE PLANE OF POLARIZATION.

"The *plane of polarization* is defined as the particular plane of incidence in which the polarized light is most copiously reflected."

When a ray of light is incident at the angle of polarization on one mirror and the reflected ray is received on a second mirror at the polarizing angle, the ray is most copiously reflected from the second mirror when the two mirrors are parallel. In this case the plane of reflection coincides with the plane of polarization. The *vibrations* of plane-polarized light are, however, in a direction at right angles to the plane defined as the plane of polarization.

(Wood, *Physical Optics,* p. 233.)

ELLIPTICAL POLARIZATION.

Light may be plane, circularly or elliptically polarized. Plane and circular polarization may be treated as special cases of elliptical polarization. Circular polarization results from the simultaneous presence at a point of two rectangular vibrations of the same period but differing in phase by a quarter period. In elliptical polarization both the phase and amplitude of the vibrations differ.

(Wood, *Physical Optics,* Chap, XI, p. 266 *et seq.* Also Preston, *The Theory of Light,* pp. 417, 418.)

ROTATION OF PLANE OF POLARIZATION.

Rotation of the plane of polarization occurs when plane-polarized light is transmitted through quartz in the direction of its optic axis, and this property is also possessed by many other substances, including many liquids and vapors. Some rotate the plane to the right (looking along the direction of propagation of the light) and are called "Dextrogyrate," some to the left and are called "Levogyrate."

(Preston, *The Theory of Light,* p. 425. Also *Ganot's Physics, art.* 687.)

BIOT'S LAWS.

The amount of rotation of plane-polarized light passed through a quartz crystal or other rotating substance is proportional to the thickness traversed by the ray. The rotation effected by two plates is the algebraic sum of the rotations produced by each separately. The rotation augments with the refrangibility of the light, and is approximately proportional to the inverse square of the wavelength.

(Preston, *The Theory of Light*, p. 426. Also Wood, *Physical Optics*, pp. 384, 385.)

POLARIZATION BY EMISSION AND DIFFUSION.

Rays emitted at an oblique angle from incandescent platinum are partially polarized perpendicularly to the plane of emission. Rays falling on a surface which is not absolutely mat, in their diffuse reflection are partially polarized.

(Chwolson, *Traité de Physique*, Vol. II, Part 8, p. 740.)

RELATION OF INTENSITY AND AMPLITUDE.

If the periods of two vibrations are the same, then the intensities of the rays are in the ratio of the squares of the amplitudes of the vibrations. Thus,

$$\frac{J}{J_1} = \frac{a^2}{a_1^2},$$

where J, J_1 are intensities and a, a_1 are amplitudes.

(On the electromagnetic theory of light, amplitude of vibration would correspond to the maximum potential-difference between opposite ends of the oscillator giving rise to the radiant energy.)

(Consult, Preston, *The Theory of Light*, pp. 43, 44.)

A NATURAL RAY REPLACED BY TWO RAYS POLARIZED AT RIGHT ANGLES.

Two rays rectilinearly polarized, which replace a natural ray, have amplitudes continually variable and, in general, unequal at each instant; the mean value of the square of the amplitude determines the quantity of radiant energy (light intensity) in the two rays.

Thus, $A^2 = \frac{1}{2}J$, where J is the intensity of the natural radiation and A^2 is the mean value of the square of the amplitude of the component rays.

(Chwolson, *Traité de Physique,* Vol. II, Part 8, p. 696.)

LAW OF LAMBERT FOR EMISSION FROM A SURFACE OF RADIANT ENERGY.

The quantity of radiant energy emitted in the unit of time, by an element of the surface of a body, in any given direction, is proportional to the cosine of the angle between this direction and the normal to the surface of the radiating body. Thus, $J\varphi = J \cos\varphi$ where J is the total quantity of energy emitted normally and $J\varphi$ the quantity emitted in the direction making the angle φ with the normal. (It is because of this law that a sphere heated to be uniformly luminous appears equally bright at the central point of its disc and at its boundary.)

(Chwolson, *Traité de Physique,* Vol. II, Part 8, p. 36.)

LAW OF ABSORPTION OF LIGHT.

When a ray enters a homogeneous medium the quantity of light of a given wave-length which is absorbed is proportional to the thickness of the medium traversed, and the amount of light which passes through a number of equal layers, diminishes in geometrical progression as the number of layers increases in arithmetical progression.

(Consult Preston, *The Theory of Light,* Chap. XVIII. See pp. 469–471. Also Wood, *Physical Optics,* Chap. XIV. See pp. 350, 351.)

DOPPLER-FIZEAU PRINCIPLE.

If a source of light and an observer approach, the frequency of the disturbance, as it passes the observer, is increased and the wave-length diminished; if they recede the reverse is true. On approach the spectrum lines are shifted toward the violet and on recession they are shifted toward the red.

(Wood, *Physical Optics,* Chap. I. See p. 19.)

HUYGENS' PRINCIPLE.

The wave-front in a train of light-waves is a surface of disturbance which results from and envelops (*i.e.,* is tangent to) the secondary waves sent out by each particle lying in the wave-front at an earlier instant.

(Wood, *Physical Optics,* Chap. II, p. 21 *et seq.* Also see Ames, *Theory of Physics,* pp. 401–403. Also Preston, *The Theory of Light,* p. 60.)

HUYGENS' PRINCIPLE OF SUPERPOSITION.

Any number of separate disturbances (light-waves) may be propagated *through* one another in the same portion of the medium. Each emerges from that portion as if it had not been encountered by others. Rays of light from all objects round about cross each other's paths in all sorts of ways, but each travels on as if the others did not exist.

(Preston, *The Theory of Light,* p. 45.)

STOKES' LAW.

The effect or intensity of an elementary wave at an external point varies as $(1 + \cos\theta)$, where θ is the obliquity, or angle between the wave-normal and the line joining the point to the center of the elementary wave. Thus the effect only vanishes for $\theta = \pi$, that is, for points directly behind the wave.

(Preston, *The Theory of Light,* p. 62.)

PRESSURE OF RADIANT ENERGY; GENERAL CASE.

When radiant energy (of any wave-length) falls normally on a *perfectly black* surface the pressure exerted on unit area is numerically equal to the total quantity of radiant energy contained in the unit of volume. If the surface is *perfectly reflecting* the pressure is twice as great. If the radiation falls on the surface making an angle φ with the normal to the surface then the pressure, per unit area is,

$$p = e\,(1 + a)\cos^2\varphi,$$

where $e = \frac{E}{V}$ is the energy in the unit of volume and a is the fraction of the energy reflected. $a = 1$ for a perfectly reflecting and $a = o$ for a perfectly black surface.

(Chwolson, *Traité de Physique,* Vol. II, Part 8, p. 84.)

LAW OF KIRCHHOFF-CLAUSIUS ON THE RELATION BETWEEN EMISSIVE POWER AND THE MEDIUM.

The emissive power of perfectly black bodies is proportional to the square of the index of refraction μ of the surrounding medium.

$$\text{Thus, } e = \mu^2 E,$$

where E is the emissive power of a perfectly black body in vacuum and e its value in any medium.

(Chwolson, *Traité de Physique,* Vol. II, Part 8, p. 83.)

DIFFRACTION OF LIGHT.

In passing through a very narrow aperture in a screen a ray of light spreads out on either side of the line of rectilinear propagation. This phenomenon is known as *diffraction*.

It is also observed under other circumstances, as when light passes the edge of a body, in which case luminous rays are bent into the shadow.

(*Ganot's Physics, arts.* 660, 661. For a detailed treatment, see Wood, *Physical Optics*, Chap. VII, pp. 150–211.)

INTERFERENCE OF LIGHT-RAYS.

When light passes through two small openings in a screen which are adjacent and the two transmitted rays meet each other at any point under a small angle the two trains of waves either annul or strengthen each other, giving dark and light bands on a screen. In the former case the distances from any point to the two openings differ by a half wave-length or a multiple of this, and in the latter case by a whole wave-length or a multiple. This reciprocal action of two wave-trains is called *interference.*

Michelson's Interferometer is an instrument for the measurement of lengths by means of the phenomena resulting from the interference of two rays of light. The instrument permits the introduction of any relative retardation between interfering-pencils of light and allows observation to be made of interference bands corresponding to a large difference of path.

(*Ganot's Physics, art.* 659. Also Preston, *The Theory of Light,* p. 202. Wood, *Physical Optics,* Chap. VIII, pp. 212–229. Also Ames, *Theory of Physics,* pp. 395–397.)

INTERFERENCE IN THIN FILMS.

If two plates of glass are put together so as to form a thin wedge of air, destructive interference of any monochromatic light will take place where the thickness of the wedge equals an even number of quarter wave-lengths and reënforcement occurs where this thickness equals an odd number of quarter wave-lengths.

(Crew, *General Physics,* pp. 502–505. For detailed account of these phenomena, see Wood, *Physical Optics,* Chap. VI, p. 100 *et seq.*)

ANOMALOUS DISPERSION.

In transparent substances in which the dispersion is normal, the refractive index increases as the wave-length decreases, but in substances which show selective absorption, the refractive index for short waves on the blue side of an absorption band is often less than the index for red light on the other side of the band. This phenomenon has been named "anomalous dispersion."

(Wood, *Physical Optics,* Chap. V. See p. 95.)

KUNDT'S LAW IN ANOMALOUS DISPERSION.

This law states that on approaching an absorption band from the red side of the spectrum the refractive index is abnormally increased by the presence of the band, while if the approach is from the blue side the index is abnormally decreased.

(Wood, *Physical Optics,* Chap. V. See p. 96.)

FRESNEL'S LAW.

"In crystals the velocities of the two light-waves are proportional to the largest and smallest radii vectors of the oval section of the wave-surface, made by a plane through the center of the surface and parallel to the wave-front."

(*New Century Dictionary, under word* LAW. For a detailed treatment of wave-velocity in crystals, consult Wood, *Physical Optics,* Chap. X. See pp. 246–257.)

NON-REVERSIBLE VISION.

If two Nicols are mounted with their directions of vibrations at an angle of 45° and between them is a medium in a magnetic field of such strength that the plane of rotation is turned 45°, light will be stopped by the second Nicol in going one way and be wholly transmitted in going the reverse way. It is thus possible to produce an arrangement whereby we can see without being seen when monochromatic light is used.

(Wood, *Physical Optics*, p. 401.)

NEWTON'S RINGS.

When a lens of large radius of curvature R is pressed upon a plane-plate of plate-glass and viewed by reflected monochromatic light, alternate dark and bright rings are produced. If t is the thickness of air-film between the lens and plate and r is the radius of a ring, then,

$$t = \frac{r^2}{2R}.$$

With sunlight the rings are many colored.

(Wood, *Physical Optics*, Chap. VI. See p. 131. Also *Ganot's Physics, art.* 664.)

THE ZEEMAN-EFFECT.

When a *source* of monochromatic radiation is placed in a strong magnetic field the lines of the spectrum are widened and many spectral lines are broken up, in the *Zeeman-effect,* into multiple lines.

(For description and explanation of the "Zeeman-effect" see Wood, *Physical Optics*, pp. 403–410.)

STOKES' LAW REGARDING FLUORESCENCE.

Stokes' law asserts that the wave-length of the light emitted by a fluorescent body always exceeds that of the exciting light.

(Wood, *Physical Optics*, Chap. XVIII. See p. 434.)

DRAPER'S LAW OF VISIBILITY.

Draper considered that all bodies begin to become visible by self-emission of light at the same temperature. He stated that this temperature is 525° C.; but the beginning of visibility is somewhat dependent upon the condition of the eye. The first visual sensation received from heated bodies is now known to be a sense of brightness to which definite color cannot be assigned. The first color-sensation received coincides with the region in the spectrum of maximum luminosity—a yellow green.

Lummer and others have shown that Draper's law does not hold true.

(*New Century Dictionary* under word LAW. See the interesting account of the supposed law of Draper in Chwolson, *Traité de Physique,* Vol. II, Part 8, pp. 33, 34 and 80, 81.)

LE CHATELIER'S LAW OF RADIATION.

The empirical law of Le Chatelier for the intensity of radiation of red light is represented by the formula,

$$I = 10^{6.7} \cdot T^{-\frac{3210}{T}},$$

where T is the absolute temperature of the radiating body and I is the intensity.

(Consult Burgess and Le Chatelier, *Measurement of High Temperatures,* p. 302.)

TALBOT'S LAW.

Helmholtz states this law thus: If any part of the retina is excited with intermittent light, recurring periodically and regularly in the same way, and if the period is sufficiently short, a continuous impression will result, which is the same as that which would result if the total light received during each period were uniformly distributed throughout the whole period.

(*Bulletin of the Bureau of Standards,* Vol. II, p. 1.)

ASTRONOMICAL ABERRATION.

The phenomenon known as the astronomical *aberration of light* is the apparent displacement of a star due to the resultant effect of the velocity of light and the motion of the earth.

This phenomenon was discovered and an explanation of it given by Bradley in 1728. The maximum displacement of a star from this cause is 20.51″ of arc.

(Preston, *The Theory of Light,* pp. 12 and 518–520. See treatment of this problem in Lorentz, *The Theory of Electrons, art.* 155. For value of the " Constant of Aberration," see *Smithsonian Physical Tables,* p. 109.)

THE ZONE-PLATE.

The *zone-plate* is a flat surface which will bring light, transmitted through it, to a focus in the manner of a convex lens. It is made by describing on a glass plate a system of circles having a common center with radii which increase proportionally to the square roots of the natural numbers, and blackening all odd numbered rings. The smaller the zones the shorter is the focal length.

(See illustration and description of this interesting optical device in Wood, *Physical Optics,* p. 31.)

THE NICOL'S PRISM.

The Nicol's prism is a most valuable device for studying polarized light. It is constructed from a rhombohedron of Iceland spar which has been cut in two along a parallel plane and the two pieces rejoined in their original position with a layer of Canada balsam or air between. It transmits the extraordinary ray only and polarizes light completely.

(*Ganot's Physics, art.* 674.)

BIBLIOGRAPHY AND INDEX

BIBLIOGRAPHY

American Journal of Science.

Ames, J. S., *Theory of Physics,* 1896.

Appell, *Traité de Mécanique Rationnelle,* 1902–1904.

Bedell and Crehore, *Alternating Currents,* 1893.

Bulletins of the Bureau of Standards.

Burgess and Le Chatelier, *Measurement of High Temperature,* 1912, 3rd Ed.

Campbell, Norman R., *Modern Electrical Theory,* 1913, 2nd Ed.

Christie, C. V., *Electrical Engineering,* 1913.

Chwolson, *Traité de Physique,* 1906–1914, five volumes.

Clausius, R., *On Heat,* edited by Hirst, 1867.

Crew, Henry, *General Physics,* 1909.

Edser, *Heat,* 1908.

Encyclopedia Britannica, 10th Ed.

Fourier, Joseph, *The Analytical Theory of Heat.* Trans. by Freeman, 1878.

Ganot's Physics, 1905, 17th Ed.

General Electric Review.

Gray, Andrew, *A Treatise on Magnetism and Electricity,* 1898.

Gulliver, G. H., *Metallic Alloys,* 1913, 2nd Ed.

Helmholtz, *Popular Scientific Lectures. Sensations of Tone.*

Hering, C., *Conversion Tables,* 1904.

Houstoun, *An Introduction to Mathematical Physics.*

Jeans, J. H., *Electricity and Magnetism,* 1908. *The Dynamical Theory of Gases,* 1904. *Theoretical Mechanics,* 1907.

Journal of The Franklin Institute.

Kaye and Laby, *Physical and Chemical Constants,* 1911.

Kimball, A. L., *College Physics,* 1912.

Lagrange, *Mécanique Analytique.*

Laplace, P. S., *Traité de Mécanique Céleste.*

Lommel, *Experimental Physics,* 1899.

Lord Kelvin, *Mathematical and Physical Papers.*

Lorentz, H. A., *The Theory of Electrons,* 1909.

Mach, E., *Science of Mechanics,* 1902.

Maxwell, James Clerk, *Theory of Heat,* 1904. *Treatise on Electricity and Magnetism,* 1881, 2nd Ed.

Merriman, M., *Treatise on Hydraulics*, 1903.

Nernst, Walter, *Theoretical Chemistry*, 1911.

New Century Dictionary.

Northrup, E. F., *Methods of Measuring Electrical Resistance*, 1912.

Planck, *Vorlesungen uber die Theorie Der Warmestrahlung.*

Poynting and Thomson, *Heat*, 1904. *Properties of Matter*, 1902. *Sound.*

Preston, Thomas, *Theory of Heat*, 1904.

Rankine, W. J. M., *The Steam Engine*, 1882.

Richardson, O. W., *The Electron Theory of Matter*, 1912.

Roscoe and Schorlemmer, *Treatise on Chemistry*, Vol. I, 1911; Vol. II, 1907.

Routh, *Elementary Rigid Dynamics*, 1905, 7th Ed.

Scientific Memoirs, edited by J. S. Ames.

Smithsonian Physical Tables, 1916, 6th Revised Ed.

Spencer, L. J., *The World's Minerals*, 1911.

Steinmetz, C. P., *Alternating Current Phenomena*, 1900.

Stokes, G. G., *Mathematical and Physical Papers.*

Thomson, J. J., *Elements of Eleotricity and Magnetism*, 1904. *Conduction of Electricity Through Gases*, 1906, 2nd Ed.

Thomson and Tait, *Treatise on Natural Philosophy*, Part I, 1886: Part II, 1883.

Transactions American Electrochemical Society.

Transactions Connecticut Academy.

Vreeland, *Maxwell's Theory and Wireless Telegraph*, 1904.

Walker, James, *Introduction to Physical Chemistry*, 1913.

Washburn, Edward W., *An Introduction to the Principles of Physical Chemistry*, 1915.

Watson, *A Text-book of Physics*, 1899, 1st Ed.

Webster, *The Dynamics of Particles and of Rigid, Elastic and Fluid Bodies*, 1904, 2nd Ed.

Wood, R. W., *Physical Optics*, 1905, 1st Ed.

INDEX

(Numbers refer to pages)

Made in the USA
Monee, IL
09 November 2026

67381880R30130